Place-Names in Glamorgan

For Marjorie

Place-Names in Glamorgan

Gwynedd O. Pierce

MERTON PRIORY PRESS

First published 2002

Published by
Merton Priory Press Ltd
67 Merthyr Road, Whitchurch
Cardiff CF14 1DD

© G.O. Pierce 2002

ISBN 1 898037 57 5

All rights reserved. No part of this publication
may be reproduced, stored in or introduced into
a retrieval system, or transmitted in any form
by any means (electronic, mechanical, photocopying,
recording otherwise) without the prior written permission
of the copyright owner.

Printed by
Dinefwr Press
Rawlings Road, Llandybie
Carmarthenshire SA18 3YD

Contents

Introduction	7
List of Abbreviations	9
Place-Names in Glamorgan	11
Further Reading	227
Index	229

Introduction

This is a selection of notes on place-names in Glamorgan which has its origin in a series of weekly articles which appeared in the *Western Mail* between 1993 and 1998. These were written in Welsh under the caption *Ditectif Geiriau* and it was a series which was inaugurated some years previously with the enthusiastic support of John Cosslett, then the Executive Editor of the *Western Mail*, the series being overseen by Rian Evans on his retirement. The original contributor was the late Professor Bedwyr Lewis Jones, and on his untimely death it was continued by the present author and two colleagues, Tomos Roberts, then the College archivist, and Professor Hywel Wyn Owen, both of the University of Wales, Bangor. The notes that had appeared between 1993 and 1996 were published under the title *Ar Draws Gwlad* (Llanrwst, 1997) and on the cessation of the appearance of the notes in the *Western Mail* in August 1998 a selection of the remaining articles was published as a second volume, *Ar Draws Gwlad 2* (Llanrwst, 1999).

These volumes were not confined in their coverage to Glamorgan but there was a preponderance of Glamorgan place-names in the whole collection which reflected the present author's own main interest. Such was the call for an English version of those notes that it was decided to prepare an English version of the Glamorgan material. To this has been added extracts from the author's writings in other sources which are not always easily accessible and permission, which is gratefully acknowledged, has been readily granted by the Glamorgan County History Trust to adapt material from the *Glamorgan County History*, Vol. II (1984) and also from the editor of *Morgannwg*, the journal of the Glamorgan History Society, to use material contributed to that journal over the years.

The random nature of the choice of material for this volume is to be explained by the fact that in some cases the notes were written at the request of those who made enquiries as a result of the appearance of the series in the *Western Mail*. The series evoked interest and a lively correspondence and was conceived with the intention that it should be read by that amorphous category, the 'general reader'.

Consequently, the presentation is in a form which some toponymists will call 'popular' in that the accepted method of presentation in place-name studies of listing forms of a name, their date of attestation together with a reference to their sources, is somewhat modified. Since this is not a formal survey of the place-names of a specific area, only the dates of quoted forms selected for the purpose (out of scores of forms collected, in some cases) and mainly the earliest of those, are cited. At the same time, there has been no attempt to avoid a discussion of aspects of philology, etymology or semantic development and although some specialist terms are employed occasionally they are kept to a minimum. In addition, a number of abbreviations have had to be used for grammatical terms, languages, a few references to frequently used sources and the names of the pre-1974 counties of Wales, in order to avoid unnecessary repetition. They are listed below.

However, since many of these notes, though not all, were written in Welsh much of what was obviously taken for granted in a Welsh reader's knowledge of the meaning of common Welsh words and terms has had to be taken into consideration and the English equivalents indicated. The remaining departure from the customary format of academic presentation in these notes is the absence of footnotes. This in no way diminishes the author's profound obligation to the work of generations of scholars in the field, from the towering figures of the past to the present, much of whose work has now achieved the status of basic knowledge. The names of the most prominent will appear frequently in these notes and a list of further reading material as a background to what is discussed in this volume is appended.

Finally I would like to acknowledge my indebtedness to friends who have favoured me with their views based on their expertise in their own fields of interest. I have had many a profitable discussion with my colleague Tomos Roberts and also with Professor Hywel Wyn Owen. Among those who are more particularly committed to their interest in the history and topography of Glamorgan who have always been ready to share their knowledge are J. Barry Davies, Jeff Childs, Richard Morgan at the Glamorgan Record Office and Miss Hilary M. Thomas.

Abbreviations

a.	*ante* (before, earlier than)
adj.	adjective
Br	British, Brythonic
Bret	Breton
cf.	confer, compare
c.	*circa* (approximately)
Co	Cornish
def.art.	the definite article
dial.	dialect, dialectal
E	English
f.	feminine
Fr	French
GPC	*Geiriadur Prifysgol Cymru* (Caerdydd, 1950–).
IE	Indo-European
Ir	Irish
Lat	Latin
m.	masculine
ME	Middle English
MW	Middle Welsh
n.	noun
OCo	Old Cornish
OE	Old English
OED	*The Oxford English Dictionary* (2nd edn, 1989).
OFr	Old French
OIr	Old Irish
ON	Old Norse
OW	Old Welsh
OS	Ordnance Survey
pers.n.	personal name
pl.	plural
PrW	Primitive Welsh
sg.	singular
vb.	verb
W	Welsh

*   before a word denotes a hypothetical or reconstructed form inferred from comparative evidence

\>   becomes, gives
\<   derived from

Aberbaedan

This name usually appears on maps as *Aberbaiden* and a recent enquirer sought confirmation, or otherwise, of the claim made by David Watkin Jones (*Dafydd Morganwg*) in his well-known history of Glamorgan (*Hanes Morganwg*, 1874) that there could be a connection between the second element of *Aberbaiden* and the *mons Badonis* or *Mynydd Baddon* 'mount Badon' which was the scene of the celebrated Arthurian victory, the *bellum Badonis* of Nennius's *Historia Brittonum*. This claim was made, one assumes, on the basis of a superficial similarity of form, for no other reason can be advanced here and the claim can be dismissed.

There is little doubt that *Aberbaedan* consists of the two elements W *aber* 'confluence of two streams', here in an inland location + the name of a stream, *Baedan,* which runs below the present *Aberbaiden Farm* (*Aber Baydan* 1633 and *Court Aberbaydan* 1664, 1731) to the north of the Cefncribwr ridge and ultimately into the Cynffig (Kenfig) river, possibly as the Nant Iorwerth Goch. Early thirteenth-century documentation is clear in its reference to *aqua de Baithan* 1199–1215, 1214–29 which appears specifically as *Nant-baydan* in 1584, and to a ford on the stream, *vado de Baithan* 1147–43 etc.

The stream, however, rises on a hill in the parish of Llangynwyd to which it has given its name, *Baithan hill,* which is *montem Baiban* (sic) 1203, *mynydd Baydan* 1695–1709, on the slope of which stood *Baiden Farm* (now demolished) in *Traen Baydan* 1578 (W *traen* < *traean* 'a third part') being one of the three 'parcels' of the parish of Llangynwyd at the time. Nearby stood a chapel of ease, *Capel Baedan* (*the chapel of Langonoyd* 1703) which has also disappeared although its existence helps to explain the reference in Griffith Jones of Llanddowror's *Welch Piety* to *Bayden* as a hamlet in 1755–6.

The stream's name varies considerably in its form over the years:

Baithan, Baythan, Baydan, Bayden, Baiden etc. but if, as is very likely, the medial *-th-* in the earlier forms attested represents the W –*dd-* (cf. the spellings *Aberthaw* for *Aberddaw(an)*, *Llantrithyd* for *Llantriddyd*) and also the probable representation of the W diphthong –*ae-* by -*ai-* or -*ay-* in local parlance, it would appear that the original Welsh form of the name was *Baeddan*. This was the opinion of the late R.J. Thomas, the foremost authority on river names in Wales. Further, the evidence of the collected forms of the name points to a provection of -*ddn-* to -*dn-* in *Baiden, Baydan* etc. which is rather unusual when there is a vowel between the two consonants, unless it has occurred under English influence.

Thomas adds that W *baeddan* was a diminutive form by the addition of the W diminutive ending -*an* to *baedd* 'pig, boar', this being an interpretation which can be accepted in view of the fact that the name can thus be put into the category of Welsh river names called after animals which rut, bore or furrow their way through the land, or in the words of Sir Ifor Williams, 'sy'n rhychu eu ffordd trwy'r tir a'r cerrig'. Such rivers are *Twrch* 'wild boar', *Banw* 'pig, piglet', cf. *Banff* in Scotland, *Gwŷs* 'pig' etc.

At the same time, R.J.Thomas also reminds us that *Baeddan* 'little boar' is attested as an old Welsh personal name so that it is possible that the stream may have been called after a person who was similarly named, in which case it falls into that other category of river names which are also personal names, like *Adda, Alun, Cynon, Dewi, Meurig* etc.

Another *Aberbaedan* (*Aberbaiden* 1803) and *Glanbaedan* (*Glanbaiden* on the modern Ordnance Survey map) exist near Gofilon, Mon.

Abergwrelych

This is where the stream *Gwrelych* flows into the river Nedd (Neath) near Pont Walby and Glyn-nedd, hence the first element, W *aber* 'confluence of two streams' here also in an inland location. Like many other stream and river names in Wales and elsewhere, being among the oldest names in existence although not necessarily attested in documentation until more recent times, *Gwrelych* is more difficult to interpret. It is a stream which rises close to Llyn Fawr in the very north of Glamorgan, near Rugos (Rhigos) and runs past the farms of

Blaen Gwrelych (W *blaen* 'source, upper reaches') and *Ystum Gwrelych* (*Estunwrelech* 1253, *Estymwereleh* 1256; W *ystum* ' bend, turn') down to *Abergwrelych*. Earlier forms of the stream name are *Wrelec* 1203, *Wrelech* 1253, *Gwrelich* 1578, *Gwrelych* 1666.

Although Sir Ifor Williams once suggested that the name could contain the W masculine ending *-ych*, Professor Melville Richards was reluctant to accept this on the grounds that what remained of the name to be analysed, the form *gwrel-*, was obscure. He had already discussed the name *Emlych* which occurs in two places, one in the parish of St David's, Pembs., and the other near Pen-bre (Pembrey), Carms., both being on the banks of a river or stream. This had led him to suggest that the second element of the form *Em-lych* was W *llych*, a variant form of *llwch* 'pool, lake, swamp' with the first element *em-* being an affected form of the prefix *am-* which is to be found in several Welsh place-names in the sense of 'around, about, beside', as in *Amgoed, Amrath* (now *Amroth*) and the Anglesey *Amlwch* (which could be connected with *Emlych,* cf. the form *Cemlyn* < *cam* + *llyn*).

Melville Richards was encouraged to accept this deduction in view of Sir Ifor's discussion of the word *gwre* (the probable first element of *gwrelych*, rather than *gwrel-*) in his seminal notes on the text of *Canu Aneirin*, namely that *gwre* (one syllable) 'small insect, mite, hand-worm' was an old singular form, pl. *gwreint, gwraint*, for which a new singular form was created, *gwreinyn*. The rest can be left to Professor Richards (in translation): ' Where the meaning is concerned, *Gwrelych* 'the grub (insect, mite etc.) stream' would conform splendidly with stream-names which contain the element *chwil* 'beetle, chafer' like *Chwil, Chwilen, Chwiler* etc.' which he had discerned in the name of the river *Wheelock* in Cheshire (cf. *Chwilog,* now a village name in Caerns.). He also added that a *Cwm Gwreinyn* occurs in the parish of Cemais, Mont. (p. 225).

Aberpennar : Mountain Ash

Most people, if asked, would probably think that of these two alternative names of this town in the Cynon valley the Welsh form Aberpennar is the older. David Watkin Jones in his *Hanes Morganwg* (1874) states that *enw cyntefig y lle oedd Aberpenar* 'the original name of the place was Aberpenar', and he was followed by others, but this is not strictly true.

Aberpennar was the name of the confluence (W *aber*) of the stream named *Pennar*, or *Pen(n)arth* in its earlier form, with the river Cynon almost half a mile upriver from the present bridge at Mountain Ash. There the Pennar ends its course from the high ground of Cefnpennar (W *cefn* 'ridge, back') on the slopes of Twyn Disgwylfa to the north: *aber pennarthe* 1570, *Aberpennarth* 1600, *Tir Aber Penarth* 1638, *Tir Aber-Penarth* 1666, *Tyr Aberpennar* 1771–81 (with W *tir* 'land'). We can accept Melville Richards's view (after R.J. Thomas) that the stream was given the name of the high ground on which it rises, W *penarth* 'height, high ground' with its variant form *pennardd* < *pen* + *garth* 'hill, height' here rather than 'enclosure' > *pen(n)ar* by loss of the final consonant in common usage. That name is not now current but it appears in Cefnpennar. John Leland referred to 'a hille ... caullid Penar' in 1536–9 which could have been a reference to this location or, more probably, to one which bears a remarkable similarity in its topography and nomenclature in Monmouthshire. The ground which rises perceptibly from the Ebwy river, on its western side, opposite Newbridge, Mon., up to Mynyddislwyn, is shown as *Penner* on Christopher and John Greenwood's map of Monmouthshire (1830) and on nineteenth-century monumental inscriptions in Mynyddislwyn church. The name now survives in the names of the farms, Pennar Farm, Pennar-fach, Pennar Ganol and Cefnpennar, all in the vicinity of a brook called Nant Pennar in that location also. As a n.f. Y Bennar occurs as the name of a prominent hill near Dolwyddelan, Caerns. and also that of a farm on sloping ground at Penmachno, Caerns.

Close to the point where the Nant Pennar runs into the Cynon on its east bank stood the house of Dyffryn (which often appears in the incorrect form Duffryn in documentation and elsewhere) the residence of the Bruce family since 1750 including Henry Austin Bruce, first lord Aberdare (1815–95). The old house was replaced by a new structure in the early 1870s and this, in turn, housed a school in 1926 before demolition in 1985–6, the location being to the rear of the present Mountain Ash Comprehensive School. As Miss Hilary M. Thomas has reminded us, the estate had been in the possession of the Jones family of Aberdare for generations and in a post-nuptual settlement of 1632–3 made by John Jones gent., the property is referred to as *Aber Pennar alias Tire y dyffryn*. This *alias* form is continued for a time: *Aberpennar alias Dyffryn* 1691, *Dyffrin alias*

Aberpennar 1717. It was also referred to as *Dyffryn Aberdare*, *Dyffryn Aberdâr*, as on George Yates's map of 1799, *Duffrin Aberdaer*, to indicate its proximity to Aberdare and as a means of distinguishing it from the scores of other places called Dyffryn in Wales. No doubt the name Dyffryn here refers to the narrow course of the Pennar brook down from Cefnpennar (W *dyffryn* 'valley, glen' < *dwfr* 'water' + *hynt* 'course, path') at the southern end of which the estate is situated, much the same as occurred a short distance away further down the Cynon valley where the Ffrwd brook (W *ffrwd* 'stream, torrent') ran down Dyffryn Ffrwd into the Cynon at Aberffrwd. This was near the demolished Ffrwd Bridge over the old Aberdare Canal and close by the present Ffrwd Crescent in Mountain Ash. In fact, Aberffrwd would have been a more accurate Welsh name for Mountain Ash than that which it eventually acquired.

An account written in 1897 states that on the opposite bank of the Cynon a few houses had been built in 1809 which included the *Mountain Ash Inn*. By 1830 a warehouse and one other building had been added excluding the surrounding farmhouses and the house of Dyffryn further up the river. By 1852 a community of about six hundred households is noted and in the following year T.W. Rammell in his report to the General Board of Health on the sanitary condition of the inhabitants of the parish of Aberdare notes that 'at the southern extremity is a village called Mountain Ash with a population of about 1,000'. Such were the early stages in the growth of the town of Mountain Ash whose name was clearly that of the inn around which the early community adhered. It follows a pattern of name-giving which is well evidenced in Nelson in the parish of Llanfabon (and its Lancashire counterpart) or Fleur-de-lis (a brewery) in Bedwellte, and Crosskeys, Mon., Holland Arms, Anglesey, Farmers and possibly Tumble, Carms. etc. In Tredegar, Mon., it was the Tredegar Iron-works which first served as the postal address of the surrounding community and, similarly, it was the community dependent upon the combined works of the Rhymney Ironworks Company who gave the town of Rhymni (Rhymney) at the northern end of the Rhymni valley its name rather than the river itself. It need hardly be added that many other places in Wales took the name of the Nonconformist chapel which was their centre of local religious and social activity—Bethel, Carmel, Bethesda, Libanus, Saron etc.

When, exactly, the name Aberpennar began to be used as the Welsh alternative for Mountain Ash has been the subject of debate.

Glanffrwd, in 1878–88, was deploring the fact that it was a 'place which had an English name, Mountain Ash' and no Welsh name. It is not necessary, here, to follow all the suggestions that have been made but it is pertinent to note that by far the most plausible is that matters were brought to a head by the visit of the National Eisteddfod to the town in August 1905. Confirmation of this fact appears to be provided by the bilingual title-page of the 120-page official programme of the Eisteddfod divided into two halves, the top half Welsh which has 'Eisteddfod Genedlaethol Freninol (sic) Cymru. Aberpennar, Awst 7, 8, 9, 10 a'r 11, 1905' to correspond with the bottom half's 'The Royal National Eisteddfod of Wales. Mountain Ash, August 7, 8, 9, 10 & 11, 1905'.

Arthur's Butts Hill

On a number of old maps of Glamorgan this is the name which is given to the Garth hill, Pen-tyrch. For example, *Arthors buttes hill* 1578 Christopher Saxton, *Arthurs Butts Hill* 1610 John Speed, *Arthur's Buts Hill* 1645 Johan Bleau and *Arthur's Butts Hill* 1729 Emanuel Bowen.

The hill has a distinctive characteristic in that on its quite level summit stands a row of four large and one smaller grass-covered mounds. From a distance they look like upturned saucers but they are the remains of Bronze Age burials.

As can be readily appreciated, much that remains of early structures, mounds, standing stones and the like were given names by the populace which were homely and familiar with an occasional attempt to associate such remains with known legendary or historical personages to whom, often, supernatural powers were granted. In Wales, it would be difficult to find a more likely recipient than the legendary Arthur. His association with ancient monuments in popular nomenclature are many: Cerrig Arthur, Carnedd Arthur (W *carnedd* 'cairn'), Ffynnon Gegin Arthur (with W *ffynnon* 'well' and *cegin* here in its old sense of 'ridge, back') are examples in Caerns., and in a number of places Coeten Arthur is the name of the supported capstone of a Stone Age burial chamber, a so-called 'cromlech', this being a diminutive form *coeten* (E *quoit* > W *coet* + the diminutive suffix –*en*), namely 'Arthur's quoit' seemingly having been hurled to its location by the supernaturally-endowed hero. Arthur's Stone corresponds to this in some areas such as Cefn-bryn, Gower, although the Welsh name in this case is Maen Ceti, W *maen* 'stone, rock' + an old personal name Ceti which is also attested in the name Sgeti (Sketty), Swansea, from an original Ynys Geti (*Enesketti* 1319, *Enesketty* 1322).

In England a term which is used for round-topped mounds similar to those on the Garth hill is *butt*. The meaning of *butt(e)* (possibly adopted from the French *butte*) was 'mound, hillock' which came to be used in the specialised sense of 'archery butt' and was sometimes used in popular names of natural hills or tumuli, again associated with heroes, in this connection none more prominent than that prototype of archers, Robin Hood. Robin Hood's Butts near Weobley, Herefordshire, are natural hills, whilst the similarly named location at Otterford, Somerset are tumuli.

It would appear that this is the pattern followed in likening the Bronze Age mounds on the Garth hill to butts used by Arthur. What may be said for the prowess of Arthur as an archer is another matter, historical accuracy in such particulars being not necessarily a feature of popular attributions.

Beaupre : Y Bewpyr

Here, at Old Beaupre as it is now called, in the parish of St Hilary, stand the ruins of the Basset mansion, the family having been connected with the vicinity since the thirteenth century. The present building, however, with its notable porch, was built in 1586.

In its French guise, Beaupre, the name has been consistently misinterpreted as being composed of the elements *beau* 'beautiful, fair' and *pré* 'field, meadow'. Whilst the first element is certainly present, the second is not what it is assumed to be.

The earliest documentary form attested hitherto is *Bewerpere* 1376, followed by a variety of forms including *Y Bewper* 1485–1515, *Beawpier* 1511, *the Bewpere* 1526, *Bowper* 1527, *the Bewpyre* 1533–8 etc., some, from both English and Welsh sources, being preceded by the definite article. It is not until well into the seventeenth century that the form Beaupre becomes common but there is nothing among the quite substantial number of those forms which have been collected to suggest that the final *-e* has a full accented vocalic quality. The local pronunciation approximates to 'Bewper'.

It is the first of the recorded forms quoted above, *Bewerpere*, which best indicates the true nature of the second element in the name. It is a compound of the OFr *beu*, *beau* 'fair, beautiful' and *repaire* 'a place of retreat' where metathesis of the first two letters of the second element has occurred in common parlance. This was followed by the loss of the 'dark' unaccented syllable *-er-* in *Bew(er)per(e)* to give the forms *Bewper*, *Bewpyr* and the like. The later form, Beaupre, has the trappings of rationalisation where the final *-re* is perhaps meant to have the semi-vocalic quality of that consonantal combination which is found in French words such as *vivre*, *prendre* etc. as a representation of the indistinct unaccented final *-er*, *-yr* of the *Bewper*, *Bewpyr* forms, and not the accented final

-*e* of the French word *pré* 'meadow'.

It cannot be claimed that this development is entirely due to Welsh influence because similar forms can be found as place-names in England. Beaupre occurs in the county of Huntingdon, Bewper in Surrey (*Beaurepeyr* 1296, *Bewpere* 1472) and Kent, and Beaurepaire (*Beaurepeir* 1346) in that fossilised form, in Hampshire. It is also the same name (*Beurepeir* 1231, *Beurepeyr* 1251) which is now found in the form Belper (Derbyshire).

Bedwellte

This is the correct form of the name which now appears as Bedwellty, and it is not always easy to persuade enquirers that the final syllable has no connection whatsoever with W *tŷ* 'house'. Moreover, several other attempts at an interpretation of the name are on record, of which a couple of examples can be quoted. In an eisteddfodic prize essay in 1884 Evan Powell suggested that it was a reduced form of a Welsh salutation, *byd gwell i ti!* (literally, 'may a better world be yours') whilst several other pundits declared themselves as being favourable to a reduction of W *bedwelltydd*, a contrived compound of W *bedw* 'birch tree(s)' + *elltydd*, a plural form of W *allt* 'wooded slope' in south Wales. None of these took cognisance of earlier recorded forms of the name, *Bedwellte* 1411, *Bedewellty* 1431, *Bod Mellde* and *Bodwellde c.* 1566 in an important list of Welsh parishes, and *Modwellty* in a genealogical collection by Gruffudd Hiraethog *c.* 1550.

These forms strongly suggest that the first element is W *bod* 'homestead, dwelling' as in compounds like *preswylfod*, *hafod* etc. which seems to be commoner in north Wales, as in Bodorgan, Bodewryd, Bodedern etc., than in the south, although it may well occur in Bedwenarth, Bodringallt and Watford, Caerffili (p. 213), even Bedlinog (rather than W *bedd* 'grave').

Normally, but not exclusively, W *bod* in compound place-names is followed by a personal name. It is known that Mellte occurs as the second element with W *ystrad* 'valley bottom' in the name Ystradfellte, Brecs. In that instance, however, it is the name of a river, which is *Meldou, Melltou* in the *Liber Landavensis*, but since R.J. Thomas has shown conclusively that many rivers and streams in Wales have been given personal names (p. 14) this could have happened in the case of Mellte. It is not a well-evidenced personal name, it may be granted, but one significant example occurs in an early thirteenth-century genealogical tract known as *De Situ Brecheniauc* where one of the numerous daughters of Brychan Brycheiniog is said to be buried in an unknown grave *sub petra Meltheu* 'under the stone of Meltheu' (a scribal form of Mellteu).

It is reasonable, therefore, to consider the personal name Mellteu > Mellte in common parlance, as the second element of the name Bedwellte. If the first element is *bod*, the initial consonant is lenited > *Bod-fellte* with the subsequent known colloquial tendency in Welsh to interchange *f* and *w* (as in *cafod/cawod*) > *Bod-wellte*. The variant form in *Bed-* > Bedwellte seems to be a feature of *bod-* names in south Wales, as noted above, and the modern form Bedwellty may have evolved from a haphazard pronunciation of the final unaccented vowel and, possibly, by analogy with the very common W *tŷ* 'house'. It is recorded in this form as early as 1584.

Blaenegel

On the modern Ordnance Survey map the name of this farm which stands in the wide hilly expanse of the northern part of the parish of Llangiwg is Blaenegel Fawr (with W *mawr* 'great, greater'), a little south of Gwauncaegurwen and Cwmllynfell. As the names of the neighbouring farms, Fforchegel (with W *fforch* 'fork') and Rhydyregel (with W *rhyd* 'ford') imply, the second element *-egel* is a river or stream name, all these farms being in the vicinity of its 'upper reaches', which is the significance of W *blaen* in *Blaenegel*.

The name Egel was discussed by Professor Melville Richards over twenty-five years ago in a note on Regal, the name of a farm in Betws Garmon, Caerns.. He was of the opinion that it was a reduced form of *Yr Egal*, the typical Caernarfonshire variation of Yr Egel, with the W def.art. *yr*. Later, Professor Bedwyr Lewis Jones drew attention to further examples of what appeared to be the same basic form, namely *Hegal*, *Hegyl* and *Hegla*, and suggested that the basis of these forms was the W word *hegl*, pl. *heglau* 'leg, thigh, shank' which occurs as an element in place-names in the sense of a ridge of land of that particular shape or form.

In the case of Blaenegel and the stream name Egel in the parish of Llangiwg, however, none of the forms of the names seen hitherto contain a trace of initial *h-*. A good sequence of documents relating to the Ynyscedwyn estate from 1556 to the end of the seventeenth century contain the name of the stream, the forms being *Eg(g)le*, *Egele*, *Egel(l)*, *Egel* (*Ecel* in the common speech of Cwmtawe). Also, *Blaenegl* 1578, *Blaenegell* 1687, *Blaen Egel* 1697, *Blaenegel* 1741, *Blanegal* 1772, 1783 etc.—all without the *h*-.

It is fair to conclude, therefore, that the name of the stream is the W n.f. *egel*, defined in GPC as a plant of the cyclamen family, particularly *cyclamen europaeum*, the roots of which are eaten by swine: *sowbread*. This is to be found growing wild in woods and thickets. Such a name can be placed in the category of other river or stream names which are those of plants which commonly grew on their banks, like Craf (Abercraf, Crafnant) 'wild garlic', Castan 'chestnut', Celynen 'holly', Cerdin 'rowan, mountain ash' and Dâr 'oak'.

Bolgoed

Several examples of this name occur in Wales, three of them in Glamorgan. It is the name of a farm in the parish of Pendeulwyn (Pendoylan), *Y Bolgoed* 1585, 1681 etc. and another between Pont-lliw and Pontarddulais in the parish of Llandeilo Tal-y-bont, *Y Bolgoede* 1567, *bolgoed* 1584–5, which was divided into two holdings, *Bolgoed Isha* 1789 (W *isaf* 'lower') and Bolgoed Uchaf, *Bolgod Ycha* 1650, *Bolgoed Ycha* 1661, 1783 (W *uchaf* 'higher') . The third is on Hirwaun in the parish of Aberdâr and is recorded at an earlier date, *Bolchoyth* 1253, *Bolgoyth* 1256, *Bolgoid* 1536–9.

It is a compound of two Welsh elements, *bol* + *coed* 'wood', the pronunciation in common speech in south Wales being *bolgod* (cf. how the name Bargoed was 'restored' from the correct *bargod* on the mistaken assumption that the latter form was a colloquial rendering of the former). The first element, W *bol*, *bola*, MW *boly*, cf. Ir *bolg*, derives from a root-form which produced the Celtic *bolg*-, and is noteworthy in that it can be understood in two opposing senses, either 'depression, concavity' or 'swelling, convexity'. In Welsh it is used to refer to the 'belly, abdomen' of the human body. When used in a topographical sense it can obviously refer either to natural hollows or depressions, even a marsh or swamp in such locations, or to mounds, rounded hillocks and similar features. In the case of Bolgoed, such locations were wooded and an interesting tautological form occurs as a name with the addition of W *bryn* 'hill' in Brynbolgoed, Glyn Tarell, Brecs.

In two of the Glamorgan examples noted above, those in the parishes of Pendeulwyn and Llandeilo Tal-y-bont, *bol* in the sense of 'hill, ridge' seems appropriate. This could be true of the example on Hirwaun but an undulating, uneven surface cannot be ruled out here although, perhaps, the plural form *bolion* could convey that sense more effectively, as in the name of the Anglesey commote of Taly-bolion (the onomastic tale of the second branch of the *Mabinogi* notwithstanding) which contains two places named Rhos-y-bol and Pen-bol, the nature of its terrain being, in the opinion of my colleague Tomos Roberts, 'undulating with low mounds and depressions'.

John Leland, the antiquary, notes the Hirwaun Bolgoed in his *Itinerary* through Wales, 1536–9, and offers an English translation of the name, 'the bely of the wood'. He was, of course, not far wrong though he may not have appreciated which of the two elements in the name was the generic, but W *bol* and E *belly* are cognate. It may be as well to remember that Bellimoor is now the name of a farm near Madley, Herefordshire, but is to be found in OW orthography in the *Liber Landavensis* as *bolgros*, this being the modern *bol* + *rhos* 'moor, heath', an inversion of Rhos-y-bol, Anglesey.

Y Brombil

On turning right, travelling from Cardiff to Port Talbot on the old A48 road, opposite the well-known hostelry of The Twelve Knights at Margam, the road runs under the M4 to the small village of Brombil which is situated at the mouth of a *cwm*, Cwm y Brombil, which penetrates the hillside to the north.

In the local Welsh speech the name is usually preceded by the Welsh definite article *y*, thereby giving it the appearance of being a Welsh form which seems to conform with the pattern of names like Y Bala, Y Barri etc. This is not the case, however, although the appearance of the Welsh definite article in local parlance could well be an analogous development.

Like many of these small valleys in the area, the *cwm* was once densely wooded and isolated (there is record of *fforest y Brombil* in 1711, 1736) with an occasional farmstead dotted here and there before the search for and exploitation of minerals began to change the landscape in the nineteenth century. Eventually, there was a coal mine at the far end of the valley and a farm, Brombil, but as early as 1551 there is record of a hamlet whose name takes the form *Bromehill*. That is the key to the language and the meaning of the name, namely E *broom* + *hill* signifying a slope where broom grew and flowered in sufficient quantity as to characterise the location to the extent of conferring upon it its name.

A simple English descriptive name, therefore, possibly one of the commonest to be found in Wales and England, the Welsh equivalent being Bryn Banadl (W *bryn* 'hill' and *banadl*, *banal* 'broom') or Bryn Banal between Pontyberem and Llan-non, Carms., Cae bana(d)l, Banhalog, Y Fanhadlog (an adjectival form in -*og* to signify 'a place where broom grows') being also common (p. 74).

The medial -*b*- in Brombil is an English development. It had already appeared in *Brombell* 1540, *Brombil* 1543, *Brombile* 1544, where -*m*- is followed by a vowel (in this case after loss of -*h*- in spoken forms) as is seen in the E *bramble*, OE *brēmel* which, as it happens, derives from a root-form which it has in common with *broom*.

To the east of Brombil stood the small village of Y Groes, 'the cross', with its notable octagonal Nonconformist chapel which now stands in Tollgate Road, Port Talbot, since its removal as a consequence of the construction of the M4 motorway. It is to that location that *Cross-Brombil* 1756–7 refers in the register of the circulating schools of Griffith Jones of Llanddowror, and it is *Brombel Cross* which appears on Emanuel Bowen's map in 1729.

The brook which ran down the small valley is Brombil Brook on the present Ordnance Survey map, but at a much earlier date its name was *Yr Annell* (a derived form of *Eiriannell*, see p. 204) a vestigial form of which can now be seen in the name of Rhanallt Street in Taibach.

Brynygynnen

This name is no longer current but its significance lies in the fact that the present building which, as a structure, perpetuates a connection with the one which originally bore the name is that which is effectively the core of the Cathedral School at Llandaf, Cardiff. This was erected as a new and spacious Georgian residence between 1744 and 1751 for Admiral Thomas Mathew, probably the best known member of the Mathew family of Radur and Llandaf, but it was a replacement for an older manor house of the family which had stood on a low ridge to the south-east of the cathedral itself and its attached Old Bishop's Palace or Castle.

This earlier house had been completed by the end of the fifteenth century, reputedly by Sir David Mathew whose family had secured the possession of the entire manor of Llandaf by 1553. In 1536–9 John Leland noted the house without giving it a name but added the comment that it was 'welle buildid', and he was followed in a tract or 'Breviat' of the history of Glamorgan in 1596 by Rice Lewis who said that it was a *lardge Demeasne which butteth to the River of Taffe* but again omitting to give it a name. It was left to Rice Merrick of Cottrel to supply this information in 1578 and it is he who tells us that it was *Bryn y Gynnen*. This is confirmed in an account written after its demolition, *Bryn y gynnen* 1752 and an undated genealogical note gives the name as *Bryncenyn*. The old house is depicted on an inset of Llandaf on John Speed's map of Glamorgan in 1610.

The new house was later known as Llandaff Court, then the residence of Miles Mathew, and by 1850 it had become Bishop's Court, the residence of the bishop of Llandaf (Alfred Ollivant at the time), and having had a small chapel added to it. Frequent references occur also to it by the Welsh title of Llys Esgob. However, after a fire in 1914 the roof and upper floor had to be reconstructed and by 1939 it had ceased to be the bishop's residence. It was further enlarged and the Cathedral School was transferred there from premises to the west of the cathedral in 1958.

As will be gathered from what has been said above, references to the old name of Brynygynnen are not plentiful, indeed, they are fewer than one would wish for the purpose of making a firm statement of belief about its original form. It has to be taken at its face value but as such it does fall into a recognisable category of names. As it stands it is a compound of W *bryn* 'hill' + W *cynnen* 'dispute, discord, dissention', a borrowing from the Latin *contendo*, the reference being probably to land or property which had once been a matter of dispute and contention. It conforms with a number of names which record similar difficulties such as Rhos Ymryson (*rhos* + W *ymryson* 'contention, conflict'), Bryn Dadlau, Clun(y)dadlau (containing the W verb *dadlau* 'to argue, dispute') with which can be coupled Trehwbwb and Cwmwbwb (pp. 198-9).

Caerdydd : Cardiff

It is now known that a sequence of four overlapping forts, from the first Neronian structure of c. 50–80/90 AD to that of the late third-century or early fourth-century Roman stone fort, with the quadrangular outlines of which we are more familiar today, occupied an area on the east bank of the Taf in Cardiff. As there is no distinct Romano-British name recorded in the classical sources for such an establishment it is logical to accept that a compound locative name, of native origin, could have been formed to signify 'the fort on the Taf', or one that implies a genitival connection with the river. That name was the progenitor of the Welsh form *Caerdyf*.

Crucial to the understanding of the structure of this name is an appreciation of the fact that it must have been formed in the Romano-British period. Although there can be no means of knowing precisely to which of the early forts the name applied, it came into existence before the completion of the evolution of the Welsh language from its inflected British predecessor, a process which was completed by the middle of the sixth century, and a stage in this process was the loss of British final syllables which previously, in many cases, had caused vowel affection in preceding syllables.

The Br root-form **Tam-* is now considered to have been the source of the name of the river Taf and in its favour is that it appears to be the source of a large category of river names in Britain and on the Continent such as Tamar, Thames, Tame, Teme, Thame, Team, Tambre (Spain), Tammaro (Italy), Demer (Belgium) Tamaran (France) etc., including the W *Tafwy*, a by-form of which was *Tawy* which became Tawe. Views about its meaning have varied. It was long thought to have a base which meant 'dark' but is now thought to mean 'to flow, to melt'. The basic form would have been **Tamo-* > *Taf*, but an oblique case-ending like the genitive *-i*, as in **Tam-i-*, would give the Welsh form *Tyf* by ultimate *-i* affection.

Prefixed by a form which gave the W *caer* 'fort, stronghold, enclosure', namely the Br **kagro-* (not Lat *castra*, see pp. 80–1), the compound W form *Caerdyf* would have evolved. All the earliest recorded forms of the name in the eleventh and twelfth centuries confirm that this is precisely its original form, *Villa Cardiviae* 1081, *Kairdif* 1106, *Kayrdif* c. 1126, *Kerdif* 1133–47 etc. Its period of origin can be emphasised by contrasting Caerdyf with the name Llandaf which cannot possibly be earlier than the Early Christian Age and where there may have been an early monastic settlement by c. 680 AD. This latter is one of the few examples of W *llan* being used with a river name, in this case to indicate its location at a crossing point to the north-west of Cardiff. Its syntactical structure is identical with Caerdyf, but being of a later period of origin by which time the Welsh language had lost all traces of inflection, it conforms with the normal Welsh practice of placing the noun which is in the oblique case immediately after the noun on which it depends, without change of form. No recorded form of the name Llandaf has any hint of the vowel in the second element being any other than *-a-*.

It is clear, therefore, that what is today assumed to be the anglicised form of the name, Cardiff, essentially preserves the original Welsh form, Caerdyf. The form *Cardif* is recorded at least as early as 1147–83 and *Cardiff* 1477. The main features of anglicisation, if they are such, are the pronounciation of the final W *-f* (E *-v*) like the E *-f* (W *-ff*) and written in the accentuated form *-ff* with a subsequent loss of the preceding long vowel quality, just as the river name *Taf* itself became Taff, similarly *Llandaf* > Llandaff. In *Cárdiff* and *Llándaff* also the accentuation has moved firmly on to the penult whereas the Welsh pronounciation was on the ultimate syllable.

The most obvious change in the form of the name which has occurred is that which is seen in the present accepted Welsh form *Caerdydd*. Some may have seen here in the form taken by the second element an analogy with the common W n.m. *dydd* 'day' but while this may have helped to perpetuate the form after the change from *-f* to *-dd* had occurred, it has no relevance. What is often disregarded by those who would arrive at such conclusions is that far more often than not, before the days of widespread literacy, changes which occur in the forms of names are phonological, or sound changes which occur in common usage and not so much changes wrought by semantic considerations. The form Caerdydd is a case in point. The interchange of *dd* and *f* in final and medial positions is an attested

Welsh colloquial change, as in W plwyf > *plwydd*, nwyf > *nwydd*, nof 'sap, juice' > *nodd*, Eifionydd > *Eiddionydd*, goferbyn > *godderbyn* etc.

Furthermore, this form of the name cannot be attested, as far as is known at present, any earlier than the sixteenth century, *Cardithe* 1553–5 (where *-th-* = *-dd-*), *(c)aer dydd c.* 1566, *Cardith* 1600–7, with *(c)aerdydd* and *(c)aerdyf* both occurring in an early seventeenth-century source.

Caerffili : Senghennydd

No surviving datable forms of the name Caerffili can be evidenced before the building of Gilbert de Clare's first castle in 1268. No direct documentary evidence is available before 1271 and in one of a number of documents of that year the castle is specifically *castrum de Kaerfili*, in another it is stated to be situated *juxta Kaerfili* 'near Caerffili'. This suggests that the name predates de Clare's edifice, being probably that of a location which was the centre of administration of the Welsh commote of Senghennydd Is Caeach. Professor William Rees was of the opinion that the commotal hall or courthouse had been situated 'not improbably on the site, and in continuation, of the Roman fort', the latter being situated to the north-west of the present ruins of de Clare's huge second castle, the so-called 'redoubt', within the outlines of which it is also thought the castle of 1268–70 was situated. This may well explain W *caer* 'fort, fortress', often applied to Roman defensive structures (but not invariably, see pp. 80–2), as the first element in the name, but the identity of the person whose name has been assumed to form the second element, *-ffili*, continues to be obscure despite the hagiographical tradition which claims that he was the son of Cenydd, an association which helps further to sustain the erroneous theory that it is his name which forms the basis of the commote name, Senghennydd.

It is claimed that place-names in Brittany where Cenydd and his father Gildas are considered to have been active in the early Christian period contain Ffili as a composite element, but this implied consequential relationship of Gildas and Ffili is itself not without its own problems of veracity and it may well be deemed inappropriate that such a person should have had a Roman fortress named after him.

On the other hand it is clear that there was a 'saint' of this name.

As Oliver Padel has shown (in *A Popular Dictionary of Cornish Place-Names* (1988), pp. 138–9), he is the patron saint of Philleigh in Cornwall (*Sanctus Filius* of *Eglosros* 1312, *Fily* 1450, *Phillie* 1613), and his name is contained, as Filii, in a tenth-century list of Cornish parochial saints, but 'of whom nothing is known' is Padel's judgement, and although he accepts the existence of the name in Wales and Brittany, 'no church dedications (to him) occur, except for one or two uncertain instances in Brittany'. Even in the case of Philleigh in Cornwall, as the first attested form quoted above shows, the alternative name was *Eglosros* (Corn *eglos* = W *eglwys* 'church' + Corn *ros* = W *rhos* 'moorland, heath') Roseland being a district of four parishes of which Philleigh is one, but there is no evidence that it ever served as 'the church of Roseland'.

The uncertainty which pervades any discussion of the second element of the name Caerffili remains, therefore, but the influence of the assumed father and son relationship of Ffili and Cenydd continues to influence various attempts to interpret the commote name Senghennydd. Cenydd's popularity in the Caerffili area remains undiminished, to judge from local nomenclature—Bryncenydd, Trecenydd, St Cenydd Road, St Cenydd Close. The local comprehensive school bears his name and the Senghennydd rugby football team is known to its faithful supporters as 'the Saints'.

Senghennydd may have been the name of a cantref (possibly the Cantref Breiniol) in Morgannwg which was divided into two commotes, Senghennydd Uwch Caeach and Senghennydd Is Caeach (that is, 'above' and 'below' the stream Caeach which served as a boundary). Much later, in the nineteenth century, the name was adopted as that of the mining community near Abertridwr, the scene of the disaster at the Universal Colliery in 1913. Disagreement about the interpretation of the name centres on the view that Senghennydd is a cymricised form of the English Saint Cenydd. '*St Cenydd* became corrupted to *Senghennydd*' is the view of one respected local historian with no other reason than that of a superficial similarity of form, it would appear, together with the prevailing assumption concerning the relationship of Cenydd and the uncertain Ffili who is assumed to have his place in the name of the commotal centre, Caerffili.

There are a number of weaknesses in this line of reasoning which cannot be fully rehearsed here, but this much can be said. Senghennydd is a Welsh name of considerable antiquity and is formed on a pattern similar to the structure of other early, and major Welsh territorial names. Melville Richards once categorised such names as names which are formed by adding so-called 'territorial' suffixes to personal names to indicate ownership of territory by the named person and his descendants. Such suffixes are *-wg* in Morgannwg 'the land of Morgan', *-ing* with *Glywys* > Glywysing, *-iog* with *Gwynllyw* > *Gwynllywiog* which became Gwynllŵg (unwisely corrupted to Wentloog under the influence of *Gwent*) etc. In the case of Senghennydd, attention should be paid to the common W territorial suffix *–ydd*, as in Eifionydd, Meirionnydd, Gwynionydd (a commote of Ceredigion), Llebenydd (a commote of Gwent-is-coed), Maelienydd (a cantref of Powys), Serwynydd (a district of old Glywysing), and *Gwrinydd* (a cantref of the old kingdom of Morgannwg named after *Gwrin* or *Gwrai*, later erroneously called Gorfynydd, Gorwenydd and Gronedd (Groneath) the centre of which was Lisworney; see pp. 121–2). More particularly, the name of another commote in Ceredigion should be noted. This is Mefenydd, where the personal name Mafan has had the suffix *-ydd* added but where the resulting accent shift has caused vowel mutation over two syllables, causing *a-a* to become *e-e*, > Mefenydd.

This is a formidable body of comparative evidence, the names being of such antiquity that in most cases the persons commemorated therein are otherwise unknown. It would be entirely consistent with the structure of the forms noted above, and with the vowel mutation over two syllables in Mefenydd, to expect a personal name, *Sangan* (for which, admittedly, there is no written evidence), to undergo a similar change of form by the addition of the suffix *-ydd* > *Sengenydd* > Senghennydd (the medial *-h-* developing normally in Welsh under the strong accent on the penult, and with the doubling of the *n* also under the accent in modern Welsh). Furthermore, the argument that *Saint Cenydd* is the basis of Senghennydd becomes less tenable when it is recalled (quite apart from the fact that it is not very likely that E *saint* or even the W borrowed forms *san(t)*, *sain(t)*, *(seint)* appear in an early toponymical context of this nature) that W.J. Gruffydd pointed out years ago that the use of the title *sain(t)* before the names of Celtic saints was neither a Welsh nor a Celtic trait. It was the Roman Catholic church which initiated the process of canonization, and although the forms *sant*, *sain(t)* etc. do appear in texts, none of these are of a provenance earlier than the twelfth century. Such forms are indicative of an acquired 'learned' mode of address and not a vernacular characteristic, much as is the fashion to add W *sant* 'saint' after the saint's name in Welsh. The custom of the Celtic church was to use the personal names of holy men without any attached title: Teilo, Beuno, Padarn etc. Cenydd would not have been an exception.

Some forms like *Seynhenith* 1307, *Seint Cenydd* early 14th c., *Seng'th alias St Heneth* 1578, *St Heineth Subtus* and *St geneth supra and subtus* 1596–1600, have been quoted in support of the St Cenydd claim. It will be noted, however, that they are all post-12th c. forms. They also emanate from non-Welsh documentary or antiquarian sources and seem obviously contrived to suggest a meaning for an otherwise 'difficult' Welsh name. They are a mere handful when contrasted with the scores of other forms evidenced which, though they may well come from similar sources and may even be somewhat distorted, are otherwise genuine attempts to represent the sound of the name without resort to the substitution of *saint* for the first syllable.

Further, the assumed connection of the name Senghennydd with the *Seinhenyd* or *Sain Henydd* of the *Brut y Tywysogyon*, which refers to Swansea and its vicinity, reinforced by the location of Llangynydd (Cenydd) in Gower, has added a further dimension to the St Cenydd saga. However, the unsolved problem of the interpretation of the Brut forms would seem to sound a note of warning against adducing them at present as proof of any related theory concerning Senghennydd (cf. p. 184).

Cae'rarfau

The name appears as Caeyrarfau, with the full form of the W def.art. *yr*, on Ordnance Survey maps. It is now the name of a substantial residence, earlier a farmhouse, on the northern edge of the village of Creigiau in the parish of Pen-tyrch. In one of the walls of the earlier portion of the house is a stone with the date 1672 cut into it.

There seems to be little doubt that the modern form of the name owes its structure to popular etymology because the second element, as it stands, is W *arfau* 'arms, weapons', the pl. form of *arf*, and coupled as a qualifier with the generic W *cae* 'field' it is not difficult for the popular imagination to conjure up the scene of a battle, a 'field of arms'.

As far as is known, there is no certain proof that such an event took place in that location but it is, nonetheless, noteworthy for having partly incorporated in a modern wall on the side of a lane which leads to the house the remains of a Stone Age burial chamber. They consist of a pair of standing stones and a cover-slab supported by one of these stones and the modern wall. It makes an open chamber which was later lime-washed and used to store coal, as we are informed by the relevant volume of the Glamorgan Inventory of the Royal Commission on Ancient and Historical Monuments. The existence of such an ancient structure, however illogically, no doubt reinforced the popular belief in this being the site of ' ... old, unhappy, far-off things, And battles long ago'.

However, the fact which cannot be ignored is that none of the earlier attested forms of the name have any hint of *arfau* as its second element. They are: *Caer Yrfa* 1570, *Karyrva*, *Kaeryrva* 1639–40, *Cae'rurva* 1650, *Kaer Erva* 1666, *Kaer Yrva* 1689, *Caryrva* 1735 etc. The evidence is entirely in favour of Cae'ryrfa as the original form, with -*yrfa*, the lenited form of W *gyrfa*, and not -*arfa(u)* as the second element. Names like Tiryryrfa and Rhosyryrfa in the parish of Gelli-gaer, Pantyryrfa in Henllys, Mon., Llannerchyryrfa in Llanfihangel Abergwesyn, Brecs., and, possibly, the name of the Maes-yr-yrfa school in Cefneithin, Carms., appear to contain the same element.

The real problem is to ascertain the meaning of W *gyrfa* in this context. In common usage it means 'course, span, career' (the 'course' which the strong man runs in Psalm 19 is *gyrfa* in the Welsh Bible—*fel cawr i redeg gyrfa*). It is also applied to a 'race, racecourse'. Such senses do not seem to be appropriate in names which have *tir*, *rhos*, *pant*, *llannerch* (as above) or *cae* as generics. Perhaps more appropriate in the case of the latter, as in the name being discussed here, would be the sense in which the word *gyrfa* occurs in one version of the old Welsh laws, namely 'a flock, herd, drove' of animals. This is echoed by Edward James in one of his homilies (1606) where he speaks (in the plural) of *gyrfâu o ychen a gwartheg* 'herds of oxen and cattle'. Used in this sense with *cae* the meaning of Cae'ryrfa would present no problem and has the additional merit of simplicity, which is the underlying principle of the process of naming. It is the lack of surviving evidence which makes interpretation difficult.

Carn Llechart

This is the name given to an imposing cairn-circle in Rhyndwyglydach composed of a ring of 25 stone slabs with a level stony interior at the centre of which is a rectangular cist whose eastern slab and capstone are missing. The W n.f. *carn* is the most commonly used element in the place nomenclature of Glamorgan to signify a cairn, mound or barrow with *carnau, cernydd, cerni* as its pl. forms. It is derived from the same root-form as Gaelic *carn*, Scots *cairn* which is the adopted form in English. Occasionally there occurs the derived W *carnedd*, pl. *carneddau, carneddi* as in Carneddi Llwydion, Eglwysilan, stone cairns of greyish colouring. The cairn in this instance is on level moorland at 940 ft above sea level, but the whole location is the mass of high ground extending for some distance north and south which forms the watershed between the Upper and Lower Clydach rivers, the northern portion of which, it is probably reasonable to assume, took its name from this monument and is known as Mynydd Carnllechart. It is presumed that a Bronze Age trackway led northwards from the cairn and in the vicinity of Baran chapel on its isolated location a mile and a quarter north-north-west in a direct line from Carn Llechart a further group of four or five cairns are recorded whose stones are in evidence in scattered locations, none of which would seem to have been sufficiently obvious to attract specific names.

The element *llechart* can be interpreted in two ways, being compounded of the W n.f. *llech* 'slab, stone' + W *garth* > *llech-arth* > *llechart* by dissimilation of *-rth* after *-ch* in common usage, a change evidenced in Glamorgan and Brecs. in the form *Cribart* for Cribarth, and elsewhere in Wales, *Llwydiart* for Llwydiarth, Anglesey, *Biart* for Buarth, Caerns., *Moeliart*, Mont., and *Sychart* for Sycharth, Denb. The main concern is the meaning of the element *garth*. As a n.f. it can mean 'enclosure, fold' cf. Ir *gort* 'field', Bret *garz*, Lat *hortus*, and it could be argued that *llech-arth* might have had the sense of 'a stone enclosure' which would appear to suit the remains of Carn Llechart, with W *carn* added later when the meaning of the form *llechart* had become obscure, for up to the present no earlier forms have been evidenced than *Garn-lachard* 1720, *Karn Llechart* 1722 (16c), *Carn llechart* 1757, *Carn-llecharth* 1830.

However, further evidence seems to point to a different interpretation which involves the meaning of the W epicene noun *garth* 'hill, ridge, height', on which the *carn* is located. To the south of the remains there is a spur of the high ground on the lower slopes of which are the two farms of Llechart-fawr and Llechart-fach, the former being at the head of the valley of the stream called Nant Llwydyn which separates this spur from the slopes of Mynydd Gellionnen and runs south-west to join the Lower Clydach river at Pont Llechart. It would seem that the two farms are a division of one earlier property called Llecharth: *Llycharth* 1584, *Llychard Fawr, Llychard Fach* 1650, *Leyghart vach* 1722, *Laughart vawr* 1740, *Laugartfawr* 1742, *Llachart fawr* 1846, and Rice Merrick's testimony is that it was one of 'the ancient houses of the family of Ho(wel) Melyn'. Having regard to the position of the farms it seems far more likely that Llecharth originates here and is a reference to the slopes of the high ground on which they, or the former single holding, stand with *garth* 'hill, ridge, height' as the second element of the name as in several Glamorgan names like Penarth, Mynydd-y-garth, Garth Graban, Garth Maelwg etc. In this case *llech* would occur in an adjectival sense to give a meaning like 'the stony height, ridge etc.'

On the slopes of the ridge to the west of Carn Llechart itself there are numerous scattered large boulders and stones which are clearly indicated on the Ordnance Survey map half a mile from the farms and in sufficient quantity to justify the tautologous appellation of Bryn-llecharth. In the vicinity of Llechart-fach, *Bryn Leyghart* was also evidenced in 1722.

Casllwchwr : Loughor

Attention has been called elsewhere in this volume to the naming of Romano-British fort settlements by that of the rivers on the banks of which they were situated. The river name Llwchwr may provide another example although the interpretation of such names presents formidable problems for the investigator, river names being amongst the earliest names in existence.

Two names are recorded in two early sources which have a bearing on this particular case. In the *Cosmography* of the monk of Ravenna *c*. 700 AD *Leuca* is recorded. A.L.F. Rivet and Colin Smith (*The Place-Names of Roman Britain*, 1979) identify this as the river Llwchwr, despite an element of doubt. In the earlier Antonine *Itinerary* of *c*. 300 AD there appears the form *Leucaro* (*Leucarum*) which Rivet and Smith identify as 'probably the Roman fort at Loughor,' and whilst conceding that there is probably a connection between the two forms the exact connotation of the latter is difficult to establish. It is generally agreed that *Leuca* stems from a Br root-form **leuc-* 'bright, shining, light', cf. W *llug* 'light', OIr *luach* cognate with Lat *lūx*, *lūcere*, the names of the rivers Llugwy, Anglesey, Caerns., Merioneth and the Herefordshire Lugg being also derived forms so called either because their waters had a turbulent or 'white' flow or were of a shimmering clarity. However, the Welsh form of the river name Llwchwr, probably earlier Llychwr, cf. W *llychwr* 'daylight' (with later vowel mutation *y-w-* > *w-w-* as in W *cymwd* > *cwmwd*, *dythwn* (< *dydd hwn*) > *dwthwn*), suggests a derivation from a form which had a suffix added to the root-form as is implied by the Antonine *Leucaro*. This could have been **-ar(a)-*, long recognised as another river or water suffix which occurs in river names like Tamar, Devon, Naver, Sutherland, and others on the Continent.

However, there are difficulties in explaining the W form Llychwr if both *Leuca* and *Leucarum* were used for the same river. The change of declension from an original **Leucarā* to *Leucaro* (*Leucarum*) might signify its use as the name of the fort or settlement. In addition, **Leucarā* would normally have given W *Llugar*. There is a river in Ayrshire called the Lugar which seems to be derived from an identical **Leucarā* and Sir Ifor Williams has shown that the *-ch-* in Llychwr, Llwchwr demands an original **-cc-* in a form like **Luccarā*. It can only be assumed for the present that this is possibly a British by-form of **Leucarā*.

The spelling Loughor betrays an obvious later English influence in the attempt to represent the Welsh form, cf. the pronunciation of Ir *lough* and the spelling Llandough for W Llandoch(au) in two places in the Vale of Glamorgan. A selection of forms from the twelfth century onwards is *Lohot* (*t = r*) 1176–98, *Lochor* 1208, *Loghern*, *Lochern* 1278, *Lozcharne* 1306, *Lochhor* 1336, *Logher* 1400, *Lloughour* 1469–70, the *-n* which appears in the final syllable is parasitic and did not endure. The Welsh name of the borough which developed around the site of the fort is Casllwchwr, *cas* being the contracted form of W *castell* with reference to the medieval castle, *Castelloghwr* 1543–4, *Kastell llwchwr c.* 1566, *Castle Lochour* 1578, *Castel Lughour* 1583, *Caeslogher* 1691, *Casllychwr* 1740–1, *Caeslwchwr* 1752.

Castell-y-dryw

South-west of the village of Llantriddyd in the Vale of Glamorgan there are remains of one of the more significant medieval moated sites in the county, that of a manorial residence on a rectangular 'island' within the surrounding moat, the western side of which was largely destroyed when a farmhouse was built at the south-western corner of the site, probably in the eighteenth century. This was Horseland farm, but this building is also now in total ruin.

A short distance to the south-west of Horseland, at the side of the road which runs from Llantriddyd to Llanilltud Fawr (Llantwit Major), stands another farmhouse, the name of which appears as Wren's Castle on current Ordnance Survey maps (*Wren, Castle* 1885, *Wren's Castle Farm* on the entrance gatepost). This is the exact equivalent of the Welsh *Castell-y-dryw* (W *castell* + the epicene noun *dryw* 'wren'), and it is as Castell-y-dryw that the farm is named on a later revised version of the six-inch OS map. However, on the first edition of the one-inch map in 1833 the name appears as Rushland (and it was so-named on the two-inch draft of 1813–14). Further investigation by the staff of the Royal Commission on Ancient Monuments, particularly Mr C.J. Spurgeon, has established that the form of the name was *Rissland* in 1956, the tenant of which testified to the amalgamation of Horseland and Rissland farms as a combined holding late in the nineteenth century. The conclusion must be that either Wren Castle or Castell-y-dryw (and it is not possible to establish which came first as there is no early dated record of either available to date) was applied as a name to the combined holdings of Horseland and Rissland 'and survives in error on the OS map as the name of the latter farm to the S.W. of Horseland'.

There is not much doubt that both Castell-y-dryw and Wren Castle are examples of the 'mocking' type of names which are fairly common in Wales and England and are usually given to places where the ruined vestiges of ancient buildings or defences were seen. The introductory volume to the English Place-Name Society's county surveys adds: 'bats, rats, frogs, owls and sparrows have succeeded to the vacant seats of the mighty'. Welsh names which can be included in this category are Castell Corryn (W *corryn* 'spider') and Castell y mwnws (W *mwnws* 'heap, spoil'), Llantrisant, Castell Crychydd (W *crychydd* 'heron', if not *cychydd* 'scavenger, wretch') and Castell Blaidd (W *blaidd* 'wolf') in Pembs. and Radnor, Castell-y-bwch (W *bwch* 'buck') Mon., Castell-y-geifr (W pl. *geifr* 'goats') Cards., and the like. In the present context it would appear that such a name could only be fitting for the site of the ruined manor house, or the later ruined farmhouse of Horseland.

Horseland is not particularly well attested as a name and to judge from the evidence available it is not of medieval origin. If the manor house had a specific name, it does not seem to have survived. *Ye Herseland* is recorded in 1689 and thereafter (*The*) *Horseland* 1717, 1734, 1784, 1788 etc. To the east, on a prominent ridge, is Coed Horseland 'Horseland Wood', which is similarly poorly documented. In view of the 1689 form the present writer at one time considered the possibility that the first element was a form of OE *herse* 'a top, a hill top' but its fairly late attestation does not fully support that suggestion. Indeed, it could be the common E *horse*, the full name having the appearance of an original field name.

The alternative Rushland or Rissland applied to the present farm of Wren's Castle may well prove to be a garbled rendering of Horseland in common usage, in view of the amalgamation of holdings noted above.

Cefn Betingau

The farm which bears this name stands on a ridge of land above one of the tributaries of the river Llan (formerly the Lliw Eithaf, see p. 155) in the parish of Llangyfelach near Swansea. It is the second element of the name which has prompted enquiries since the first, W *cefn* 'ridge, back' refers to its location and is self-explanatory. It is also relevant to note that the names of two other farms on the northern rim of the bordering parish of Llangiwg, Beting Uchaf and Beting Isaf (the 'higher' and the 'lower' Beting) are related forms.

Although the form *betingau* occurs generally on maps and would appear to be a Welsh form which has the common plural ending *-au*, this is found not to be the case when the earlier recorded forms of the name are taken into consideration: *Bettingveth* 16c., *Betting-va issa* and *ycha* 1650, *Keven betinge ycha* 1668–9, *Keven-bettingvey issa* 1693, (*Mellin*) *Bettingveth* 1695, 1756 (with W *melin* 'mill'). These tend to confirm that it is the W suffix *-fa* 'place' (*betingfa*) rather than the pl. *-au* which was the final element of the original name. This became indistinct in the vernacular in an unaccented final syllable and was reduced to -*e* and -*a*, this being later wrongly 'restored' as the plural ending -*au*.

The W n.m. *beting*, *beting*, is common in south Wales and in the form *batin*, *bating*, in the north. It is a borrowing from ME *bete*, ModE *beat* and the form *beating*, which occurs mainly in field names in the southern counties of England and particularly in Devon since the fourteenth century in a name like *Beatland(s)*. It is echoed by the Gower *Bettland* 1665, *Bett-lands* 1696 in Reynoldston and Oystermouth. The verbal form refers to the process of slicing turf with a breast-plough or paring-iron to be burnt so that the ash can be spread as manure (cf. pared turf). In south Wales the form *bieting* is also current (and in Powys, according to GPC) and this form may be the original form of the farm name Beting in the parish of Llangiwg. Among those recorded are *ebietyng* 1513, the earliest seen hitherto, and *y Byetting* 1547, *Tir y Byetting* 1751, *Bietting* 1791.

In Glamorgan the element is found in field names like *Kae r betting* 1630 in the parish of Llanfabon etc. and Bedwyr Lewis Jones has noted *Parc y Beting* in Llangynog, near Carmarthen, *Batinge* (which would probably have been another Batingau) in the Gyffylliog area near Rhuthun, Denb., *Y Betyn* in Llangoed, Anglesey, and *Cae'r Batin* in Llandygái, Caerns. Perhaps the most interesting is *Cae Abaty* in Dinas Mawddwy, Merioneth which appears to have W *abaty* 'abbey' as its second element but whose original form was probably *Cae Batin(g)* which became *Cae Baty(n)* and *Cae Abaty* in common parlance.

The W verb *betingo*, *batingo* and the operative noun *batingwr* are also related, and the noun *betingaib* used for the implement employed in the turf-paring operation is found in a couplet by the poet Dafydd Benwyn who lived in the second half of the sixteenth century (W *caib* 'pick, mattock; share').

Cefncelfi

It is the second element of this name, that of a farm situated on a prominent ridge in the parish of Cilybebyll, near the village of Rhos and in the vicinity of Alltwen and Pontardawe, which has been the subject of a number of enquiries, the first being clearly W *cefn* 'ridge, back'.

The name occurs in one of the celebrated series of *Englynion y*

Beddau (The Stanzas of the Graves) in the Black Book of Carmarthen which records the location of the 'graves' of three ancient heroes of whom we know nothing save their names, Cynon, Cynfael and Cynfeli in *Kewin Kelvi* (in the orthography of the Black Book) where *Kewin* (modern *cefn*) is clearly disyllabic, as is generally the case in common parlance in south Wales, with its intrusive vowel, modern *cefen*, to give what in present-day orthography would be *Cefencelfi*.

Although scholars are now of the opinion that the Black Book manuscript dates from the middle of the thirteenth century, it was the opinion of the late Professor Thomas Jones that the stanzas could be based on traditions which reach back to the ninth and tenth centuries. It was he who also explained the significance of that second element - *celfi* in the place-name.

It is not the plural form of W *celf* 'art, craft' (which is a relatively modern shortened form of W *celfyddyd*) but a singular or collective noun which means 'column, pillar, post' and has the same root-form as Ir *colba*, of similar meaning, and Lat *columna* (which gave us W *colofn*).

Two long stones stand on the ridge at Cefncelfi, each in adjacent fields, and reasonably reliable evidence has been found for the existence of three stones at one time. Even the two remaining stones have probably had their heads lopped off, as it were, where previously they may have been about five feet high. They are also similar in nature to many other long stones which can be found up and down the country although any vestige of ornament or inscription to confirm their use as early memorial stones or gravestones is lacking.

However, local tradition that such was their purpose is strong. Professor Thomas Jones suggested that it was the location of these stones which gave its name to Cefncelfi and that it is this tradition which accounts for the inclusion of the stanza naming the place in the Black Book. The suggestion may be more than merely plausible, in which case it is the collective meaning of *celfi* which should be accepted: 'the ridge of the columns' or 'the ridge of the long stones'.

Cefnydfa

Any discussion of the apocryphal tale about Wil Hopcyn and the Maid of Cefnydfa inevitably throws up speculation about the name of the unfortunate lady's home. The farm is another which stands on a prominent ridge or back, W *cefn*, but there is a general inclination to link the second element *-ydfa* with W *ŷd* 'corn, grain' and the common W suffix *-fa* (W *ma* 'place') to indicate either a place where corn was stored or where the land was fertile enough to bear plentiful crops.

Such an interpretation is not far from the truth in a general but not in a specific sense, for the element *ydfa* is not based on W *ŷd* 'corn' but on W *cnwd* 'crop' which has wider associations. More documentary source material now available attests the suggestion made to that effect more than thirty years ago by Professor Melville Richards.

The earliest dependable form is that contained in the writings of the antiquary Rice Merrick of Cottrel in 1578, *Keven y guydva* (where *-u-* is an error for *-n-*, the second element to be read as *gnydva*), although it may well be that this is the property referred to in what seems to be the truncated form *Keven y gynnyd* in a survey of the lands of the earl of Pembroke made a few years earlier in 1570.

However, seventeenth- and early eighteenth-century sources regularly contain forms like *Keven y Knydva*, *Keven gnydva*, *Keven y Knidva* etc. until the 1730s when *Kevenydva* (*Cefnydfa* in modern orthography) becomes the norm with loss of the initial consonant of the second element in the tight *-vn-kn-* (*-fn-gn-*) combination, but there is little doubt that this second element is W *cnydfa* (lenited initially as a feminine noun after the W def.art., –*gnydfa*).

Cnydfa is not listed in GPC but several adjectival forms based on the substantive *cnwd* 'crop, produce' are included, like *cnydiog*, *cnydfawr*, *cnydlon* or *cnydlawn*, *cnydiol* and the verb *cnydio*, *cnydu* 'to crop' on the basis of which Professor Richards has reasonably suggested that the form *cnydfa* was an acceptable substantive form with the suffix *-fa* to give the sense 'a fruitful, productive location'. It looks like a good example of a place-name which contains a form not otherwise attested in Welsh literary sources.

This is not the only recorded example. *Kayr Knydva* 1560 (that is, Cae'r gnydfa) occurs in Pendeulwyn (Pendoylan) in the Vale of Glamorgan and an earlier example, *Dolgnydfa* 1538 in Llangeler, Carms. Cnydfa as a simplex form occurs in the parishes of Llandinam, Mont., and Llanelli, Carms., and Gelli'r Gnydfa in the parish of Llanegryn, Merioneth. A further reference occurs to land in Bochrwd (Boughrood) parish, Brecs., Y Gnydfa, in the form *Y Kynydva* 1629, 1654, where the common intrusive vowel between consonants appears in the first syllable in the vernacular.

Celydd Ifan

In the discussion on Cefn Celfi in the parish of Cilybebyll (see p. 38) the opinion of the late Professor Thomas Jones was quoted to the effect that the collective form *celfi* in that name referred to two long stones (where there were once three) which stood on a prominent ridge or back of land, and that it was based on the old Welsh word *celf* 'pillar, column' and not the later abbreviation of the W *celfyddyd* 'art(s), craft'.

One wonders whether some analogous belief was in the the minds of mapmakers and estate surveyors when they dealt with place-name forms in Glamorgan because their form of the name which is noted here is almost invariably Celfydd Ifan from about the beginning of the eighteenth century onwards: *Kelvydd Ievan* 1703, *Celvydd Ievan* 1755, 1784, 1793, *Celfydd Ifan* on the first one-inch Ordnance Survey map of 1833 and the majority of official papers thereafter.

The original holding on the high ground which separates Cwm Llyfni from the small valley of Nant Cedfyw in the parish of Betws Gwair (Betws Tir Iarll) was divided to form two farms, the *uchaf* 'higher' and *isaf* 'lower', and there are grounds for believing that its previous name was *Llety Domos* 'Tomos's lodging' (of which there is more than one example in Glamorgan).

It is *Kellyth Jevan alias Llety Thomas* which is recorded in 1570, and then *Kelidd Jevan ysa* 1578, *Kelydd Ievan* 1630, *Celydd Ieuan* 1794 etc. with no trace of a medial -*f*- (or -*v*-) in the first element. This is the correct form, in which W *celydd* looks as if it could be a plural form with its terminal -*ydd*, but this is not so, although the redoubtable Edward Williams (Iolo Morganwg) believed that it was and formed a new singular form *câl* which does not seem to have had much circulation, if any. Even so, *câl* is included as an entry in GPC with a note to the effect that Iolo Morganwg sought to base it on Fr. *cale* 'wedge, plug'. In fact, *celydd* is a singular n.m. having the sense 'a woody shelter, a bower', that of some Ifan or Ieuan in this connection.

Cilibion

This is the name of a farm in Gower at the side of the road from Llanrhidian to Killay and Swansea and close to its junction with the road from Reynoldston over Cefn-bryn. The name is now associated with two other farms in the vicinity and with a wood plantation. In the Middle Ages it was the location of a grange of Neath Abbey.

The prominent feature of the area is its bareness and its wet, swampy nature. On the one hand the slopes of Cefn-bryn with the Broad Pool and other places whose names tell a similar story: Wernhalog, Welsh Moor, Pen-rhos, Pen-gwern etc. It comes as no surprise, therefore, to find that the first element in Cilibion is not W *cil* 'nook, corner' but W *celli* 'grove, copse, small wood' cognate with MIr *caill* 'wood'. This is implied in the first available recorded form *Kynthylibian* 1339, and confirmed by *Kellylybyan* 1535, *Kelly lybyon* 1557, *Kelly libion* 1583 etc.

Celli is a singular n.f., the initial consonant being lenited after the Welsh definite article, Y Gelli, and it can be followed by a second adjectival element to form close compounds like Y Gellidywyll (W *tywyll* 'dark'), Gellionnen (W *onnen* 'ash'), Gelli-gron (W *cron*, the feminine form of *crwn* 'circular, round') etc. with the lenited form frequently remaining in circulation after loss of the def.art. in everyday speech (as in the last two examples) this being the explanation for many Welsh place-names, though not all, beginning with *g*. In most of these forms the adjectival component is in its singular form, but there are exceptions.

Cilibion is such an example because the second element is the plural form of the adjective *gwlyb* 'wet, damp, moist', namely, in its modern form, *gwlybion* with its initial consonant *g*- also lenited to agree with the feminine *celli* (the lenition of *g*- in Welsh results in its disappearance). In modern orthography, *Celliwlybion* would be the form of the name for there is no evidence in the collected forms in this particular case of its being preceded by the def.art. In addition, the unaccented medial consonantal –*w*- was lost in common parlance, *Celli(w)lybion* > *Cellilybion, Cellilibion*.

As to why the plural form of the adjective gwlyb should appear as the second element in this case we can only point out that it is the singular form of the adjective which occurs in Gelli-wlyb near Crai, Brecs., but the name Gellihirion is to be found in the parish of Eglwysilan near Upper Boat (Glan-bad) where the plural of W *hir* 'long' is used in contrast with Gelli-hir which is a comparatively common Welsh place-name. This apparent inconsistency may well arise from the possibility that *celli* may have been historically a collective plural form which, like the old dual plural form in MW, was followed by the plural form of an adjective, either that or that it was sometiomes popularly taken to be a plural form in *-i*.

After *Celli(w)lybion* had become *Cellilibion*, the further loss of a syllable in everyday speech produced the modern form Cilibion, presented in the anglicised form Cillibion on Ordnance Survey maps.

Cimla

More enquiries were made concerning the meaning of the name of this outlying district of Neath (Castell-nedd) than of any other during the National Eisteddfod of Glyn Nedd in 1994.

As the town of Neath grew after the Middle Ages its bounds extended from the east of the Church of Llanilltud and Nant Illtud (or Ffos-y-gwŷdd) over Cefn Saeson in a broad semi-circular sweep down to Cwrt Sart and the Nedd river by the middle of the sixteenth century. Cimla was included within these bounds, indeed, *Kymne Kenol* (a variant form, *cimne*, and *kenol*, i.e. *canol* 'middle') is an alias of *Cefn Saeson* in documentation dated 1560, 1612–3, 1707 and it is a word which means 'common (land)', in this case, the land on which the burgesses had the right to graze their livestock. However, though the meaning is clear, its etymology is more uncertain and its form varies where it is found in different locations.

The form *cimle* and *ceimle* is defined in GPC with the suggestion that it is based on a core element *cim* which sustains the sense of 'common', citing Cim, Pont-y-cim, Efail-y-cim etc. among several examples of its provision in north Wales.

Where the form *cimla* is in question, it may well be derived from an original *cimne*, with its obscure suffix *-ne*, which developed in common usage as *cimdde*, *cimdda* and *cimle*, *cimla*, as in the development of W *simne* 'chimney' in the spoken forms *simdde*, *simdda* and *simle*, *simla*. The common at Llantrisant is Y Cimdda (*Kinme* 1578 being a dissimilated form of *cimne*) as was the common at Llandaf at one time (*Kimtha* 1709, *Kimdda Bach* 1730). It is also recorded near Resolven (*yr Cymla* 1681), Llangeinor (*Y Kymney Bach* 1631), Gelli-gaer (*le kymney* 1432), Llandow (*Y kymney Bagh* 1631) and *ye Cymle bach* 1653 which is an alias of the Cardiff Little Heath or *Waun Ddyfal*.

A vocalic variation of this is *Cymdda* which occurs in the Vale of Glamorgan, as in Coed y Cymdda, in Wenvoe (Gwenfô) and

Leckwith (Lecwydd), Cymdda Mawr, St Lythans (Llwyneliddon) etc. and similarly *cymla* appears for *cimla* which, with its intrusive vowel in common speech has become *cymyla* in the parish of St Brides Major (Saint-y-brid) and has been misinterpreted as the colloquial form of the plural W *cymylau* 'clouds'.

However, if the etymological suggestion of GPC is correct, why is it that in the case of Cymdda in the parish of Wenvoe its earliest recorded form appears as *Kymvey* 1669–73, 1730 which in modern orthography would be *cymfai* or *cymfei*? Do we have here the Welsh prefix *cym-*, *cyf-*, which often signifies 'co-possession', with the old form *mei-*, *mai-*, 'land, plain' with *cymfei*, *cymfai* becoming *cymfa* and *cymdda* in common usage, as with W *camfa* 'stile' > *camdda*?

There may well be some analogous development involved here which only the discovery of further evidence will clarify.

Coedarhydyglyn

The occasion of the death in February 1995 of Sir Cennydd Traherne, who had such an intense and knowledgable interest in the meaning of place-names in his native Vale of Glamorgan, increased speculation about the form and meaning of the name of his home in the parish of St Georges, now Coedarhydyglyn.

It is quite clear that this form of the name is of comparatively recent origin and, further, that it was concocted in order to make sense of a name that had lost both its significance and meaning to a later generation. Coedarhydyglyn is composed of three main Welsh elements, *coed* 'wood' + *ar hyd* 'along' + *y glyn* 'the valley, glen': 'the wood along the glen'. However, when it is realised that the old homestead to which the name was first given was situated on a ridge above the location of the later building (not built until 1830) it can hardly be claimed that the present form of the name makes much sense.

The local pronunciation of the name was always *codriglan*, for Coedriglan, with a strong accent on the second syllable and with the well-attested lengthening of the diphthong *-oe-* in *coed* to *cōd*. The form Coedriglan also appears consistently in a large variety of sources from the beginning of the nineteenth century. Only one prominent variation occurs, namely *Coedr(h)yglan*, and that form in turn has itself been subjected to a variety of interpretations. The antiquary

B.H. Malkin noted in 1803, 'Coedryglan, signifying Rye-hill-wood', and he was followed by several commentators, including some academics of the present era, who saw in this form the Welsh elements *coed* + *rhyg* 'rye' + *glan* 'high ground, ridge, hill'.

As far as the documentary evidence allows, it would appear that the original name was that of a wood in the area, the earliest form attested hitherto being *Reglines Wood* 1540, and the first element thereafter assumes slightly different forms: *Ragling, Rigeley* 1572, *Riglyn, Riglin* 1578 by Rice Merrick under the heading 'Woodes', *Raglande* 1591 and *Raglan* 1597–8.

The 1591 and 1597–8 forms encourage the belief that here we have the family surname *Raglan*. Members of the family of Sir John Raglan of Carnllwyd and Llanilltud Fawr (fl. 1521) held land in the parishes of St Nicholas, St Fagans, St Georges and Michaelston super Ely in the middle years of the sixteenth century, including Coed Raglan—this form being attested in the document of 1597–8. Coed Riglan can be accepted as a local variation in common usage.

Should Coedarhydyglyn be perpetuated? Coed Riglan would be better, but in spite of its artificiality, the former still has official approval.

Coedygores

This is a name which does not appear on recent popular maps, but it is not entirely lost as it has been retained as the name of part of the suburban area of Llanedern, Cardiff. Parts of the old homestead and farmhouse which bore the name (originally that of a substantial wood, W *coed*, which stood around the house) and which was in the possession of a branch of the numerous Morgan family, still exist refurbished and converted into a restaurant and public house, 'The Harvesters', which stands above the A48 or Eastern Avenue leading from Cardiff to the east.

One newspaper correspondent once vehemently claimed that the correct form of the name is Coed-y-groes, with W *croes* 'cross' as the final element, but there is not a single example among the collected forms of the name which favours this view. And what of the local historian who indulged in a similar whimsical rearrangement of letters so as to arrive at an English translation, 'the gorse wood'?

In the first place, for comparative purposes, it is relevant to note

Dr B.G. Charles's reference to the two names Llwyngoras and Llwyn(y)gorras which can be found in two parishes in Pembrokeshire. Though it can be argued that the second element in these names (with W *llwyn* 'grove' as first element) is the W *corres*, the feminine form of W *cor* 'dwarf', it may be more appropriate to consider an element which refers to the nature or dominant attribute of the plantation, since *llwyn* is the first element in the one form and *coed* in the other.

The earliest forms of the name seen hitherto are no earlier than the seventeenth century: *Koyde y Gorres* 1608, *Coed y Gores* 1685, *Coed-y-goras*, *Coyd-y-gores* 1702 etc. with the majority containing the suffix *-es*. This suggests the old Welsh word *gores(t)* which can have a possible connection with a wooded location.

Gores, gorest, was an adjective used to describe unenclosed land of the rough uncultivated kind, unfruitful land, also waste, but in names like Llwynygoras and Coedygores (the final *-es/-as* variation being well-attested in colloquial speech) it must have acquired a substantival quality as a masculine noun preceded by the definite article, *y*. The wood in question must have stood on land which had such characteristics at one time.

Coedymwstwr

The National Eisteddfod of 1998 at Pen-coed, near Bridgend, was held in the shadow of a hill which rises to a height of 370 ft, on the summit of which there is a univallate Iron Age hill-fort. The whole surrounding area in the parish of Coychurch is one in which one senses the existence of early communities, and to judge from the evidence of existing concentrations of early Christian memorial stones, none less flourishing than those which existed in the pre-Norman ecclesiastical context. The Welsh name of the parish, Llangrallo, itself denotes an early religious settlement (W *llan* + the 'saint's' name *Crallo*), as is emphasised by the discovery of the remains of two tenth-century crosses in the burial ground of the parish church as well as another of the eleventh century in the neighbouring parish of Coety.

Although only partially so by the present day, the whole area was once densely wooded, as surrounding place-names amply testify. The parish was given the name of Coychurch in post-Norman times (W *coed* 'wood' + E *church*) and in addition to the contiguous Coety, other 'wood' names in the area are Pen-coed, Tor-coed, Coed-y-gaer, Coedypebyll, Prysg (W *prysg* 'grove, coppice'), and grouped on the sides of the aforementioned hill were three farms which bore the names Coedymwstwr Ganol (W *canol* 'middle'), Coedymwstwr Uchaf ('higher') and Coedymwstwr Isaf ('lower', or earlier Coedymwstwr Bach 'the lesser, smaller') and the mansion house Plas Coedymwstwr (*Great Coed Mwstwr* 1807) which was refurbished by A.J. Williams, Member of Parliament for South Glamorgan 1885–95, now a well-known hostelry but formerly the centre of the original holding before subdivision.

The significant element in Coedymwstwr, however, is the second, - *mwstwr* preceded by the vocalic form of the W def.art. *y*. On the basis of comparative evidence it can be accepted that *mwstwr* here, through an intermediate form *mystwr*, is a colloquial form of the old Welsh word *mystwyr* which is no longer current. This is derived ultimately from Lat *monasterium*, but not directly as its form suggests that it comes from the Vulgar Latin form *mon'sterium* in which syncope of the second syllable had already occurred in the way which accounts for Welsh words like *elfen* < VLat *el'menta* < Lat *elementa* or W *mynwent* 'cemetery' < VLat *mon'menta* < Lat *monumenta* etc. and could have been borrowed as late as the sixth century. With W *mystwyr* can be compared E *minster*, Fr *moustier, moutier*, adapted into Bret as *moustoer*, and OIr *monister*, Ir *mainister*, which seems to have been derived directly from Lat *monasterium*, this being, of course, the Latin base of the later E borrowing *monastery*. It should be emphasised that the kind of establishment represented by the term in the Early Christian period was a small monastic cell and not a large edifice of medieval proportions. It is not entirely implausible to think of the original settlement of Llangrallo in such terms, but it is not possible to claim any such connection on the basis of surviving evidence. Neither is there a clue to its identification in any other circumstance.

One interesting reference occurs in the name of one of the fields of Coedymwstwr Ganol on a Dunraven estate map of 1778. This appears as *Cae'r ffunnon* (with W *ffynnon* 'spring, well': 'the well field'), the spring itself being also marked and named *Funnon-y-Munalog*. This is a non-Welsh scribe's attempt to represent Ffynnon y Fynachlog 'the monastery well', W *mynachlog* being a later compound form from W *mynach* 'monk' and *llog* < Lat *locus* 'place, consecrated place' (cf. OE *stōw*) and now (since the thirteenth century at least) the word used for 'monastery', but we can still only ask, what monastery?

Despite the uncertainty concerning the identification of the early religious settlement to which *mystwyr* may refer in Coedymwstwr, the evidence for the use of the term as an element in place-names in other locations is gradually becoming apparent. Two examples will suffice at this juncture, the first being the earliest recorded to date and which appears in the *Liber Landavensis* in relation to Llandenni, in Gwent, originally Mathenni, the reference being to *Mathenni mystwyr mawr* (in modern orthography), and where *mystwyr mawr* possibly occurs in an adjectival capacity to signify 'the great monastery'. The second example, which survives, is one where it is virtually certain that *mystwyr* was the original form of the name. This is the name of the township of Mwstwr in the parish of Corwen, Merioneth, once the property of Valle Crucis Abbey. It appears as *Mystwyr* in the abbey's foundation charter, 1222. It is further exceptionally well attested, all the forms of the name more or less containing the essential proof of authenticity, that of the existence of the diphthong -*wy*- in the second syllable < Lat –*ē*-. The final element of Coedymwstwr is nowhere near to being so well evidenced. No forms earlier than the sixteenth century have been seen: *Coide Muster* 1536–9, *Coyt mwstwr*, *Koed must(er)*, *Coyd mustwr* 1578, *Coedmustur* 1612–13, *Coedmuster* 1629, 1676, 1736, *Coed Mwstwr* 1711, 1778, 1833 etc., but the colloquial modification of the form *mystwyr* to *mwstwr* in the case of the Corwen Mwstwr together with some circumstantial evidence makes it very likely that we have here a similar appellation.

It is not W *mwstwr* 'muster, gathering' (borrowed from OFr *moustre*) or 'noise, clamour etc' as has been claimed, therefore, but a hint in nomenclature of the existence of an Early Christian religious cell which cannot yet be identified.

Cold Knap

This is the well-known name of the headland on the western side of the old Barry harbour sometimes referred to tautologically as Cold Knap Point on maps, the farm which stood on the leeward side of the headland is Cold Knap Farm on the 1885 six-inch Ordnance Survey map. However, persistence in using the epithet *cold* is not welcome and local residents will refer to The Knap more often than not, as befits a location which emphasises its attraction to visitors.

It is probably an English name in origin and not one of Scandinavian provenance as was suggested in the 1920s on the basis of the belief that the form *knap* derived from the ON *nabbr* 'headland, knoll' or ON *knappr* 'round hill', since there is a marked lack of early forms to corroborate such a claim. It is assumed here that it is E *knap*, OE *cnaepp* 'hill, top, hillock' which is also the meaning when borrowed into Welsh as *cnap*, in names like Cnap Coch, Llansamlet, Cnap Llwyd, Llangyfelach etc.

Although *Colde Knapp* is recorded in 1622 there is substantial evidence to question the validity of *cold* as the original first element of the name. The name of the farm on the knap is also recorded in 1622 and appears as *The Coale*. Thereafter it is *The Cold Ffarme* 17c., *The Cole Farm* 1705, 1713, 1784, 1811, *Coalstone* 1727, *Ye Coal Ffarm* 1789, *Cole farm* 1815, and a form which appears in the hand of Iolo Morganwg in 1786, *Côl, wrth Barri* ('Côl, near Barry'). *Cold* appears only once in this list, the variations *coal(e)*, *cole* are predominant and are echoed in his own orthography by Iolo Morganwg who usually based his written forms on the spoken forms as he heard them.

In the 1622 survey quoted above a reference is also made to *Colehole*, but without an indication as to the exact locality save that it was in the vicinity of *The Coale*. Further, on a Wenvoe estate map of 1762 the farm is certainly named *Coal*, in the immediate vicinity of which are marked several *coal-pits*. The evidence is sufficiently strong, therefore, to accept *coal*, OE *col* 'charcoal', which assumes the forms *cole* and the more usual *coal(e)* as the original first element of the name of the farm. The references to *Colehole* and *coal-pits* are clearly to charcoal-burning 'pits' or hearths, often depressions in the ground which did not admit air, so that rapid combustion was prevented in the process of charring, and not to deeply sunk pits of a later era.

It is thus reasonably clear that the first element in Cold Knap is the same in origin as that found in Coal Farm, the headland forming, as it seems to have done, a fair proportion of the land which was attached to the farm—what is generally termed in the 1622 survey as *The Coale Lands*. The change to *cold* can be attributed to popular etymology as being descriptive of a headland jutting out to sea, when it acquired that name by association with Coal Farm, rather than as the name of the comparatively sheltered farmstead on its leeward side.

Corntown : Cortwn

> Gwenni fach o Gortwn,
> Sant-y-brid a Nortwn
> Ffontygari, Sant Hilari,
> Ffonmon a Silstwn.

This is one of those popular Glamorgan Welsh folk verses known as a *triban* which in this case is composed mainly of the names of places in the Vale. It may be paraphrased as follows (using the English forms of the names): little Gwenni from Corntown, Saint Brides and Norton, Fontygary, Saint Hillary, Fonmon and Gileston. In the Welsh version the main rhyme is carried by the suffix *-twn* which is the particular form of the Welsh rendering of E *-ton* in south Wales, *Nortwn* being for Norton and *Silstwn* for Gileston (as compared with *-tyn* in the north-east, as in Mostyn, Mertyn, Prestatyn). Similarly, *Cortwn* is the Welsh form of Corntown, near Ewenni, and not *Corntwn* as has been suggested recently, the *-n-* being lost in the combination *-rn-* in common parlance as in W *tyrpeg* for E *turnpike*.

The English form Corntown has been wrongly interpreted as consisting of the elements *corn* + *-ton* 'farmstead, homestead' (> *–town*), but although this can be stated quite confidently it is difficult to be equally positive in suggesting an alternative. One thing is certain, namely that the final element was not the common E *-ton* originally but rather the OE *dūn* 'hill', ModE *down*. Further, the available evidence implies that the first element was not originally monosyllabic, *corn*, but disyllabic. The early forms are: *Corundone*, *Corendone* 1226–9, *Corend(on)* c. 1260, *Corndune* 1262, *Coryndown* 1459. In 1459 *Corntoun* appears and remains thereafter with or without slight modification.

If the first element is English, it is difficult to determine its meaning, but in a linguistically mixed area it would not be unreasonable to find a possible Welsh alternative which could produce a hybrid form of Welsh/English components such as Brynhill, Bryndown, Penhill etc. which can be found in the Vale of Glamorgan.

W *corun* 'summit, head, crown' < Lat *corōna* has been suggested as a possibility, and is worthy of further consideration. The name Corringdon in Devon was *Correndon* 1284, *Corndon* 1288, 1292 where there is a rounded hill, *Correndon Ball* 1575, and the cone-shaped Corndon Hill in the parish of Hyssington in the old Montgomeryshire was *Corendon* 1275, *Korundon* 1378. Both look like hybrid forms but the latter was cymricised as *Cornatyn* which contrasts, perhaps significantly, with the Glamorgan *Cortwn*. Whatever the discovery of further early testiomony may reveal, if it exists at all, we can at least put aside any idea of a connection with cereals of any kind, even at this uncertain stage.

Craigybwldan

A short distance to the south of Waunarlwydd and on the slope of the ridge which extends westwards from Cocket, on the western fringes of Swansea, and which explains Cefn-coed as an element in several names in the vicinity (W *cefn* 'ridge, back'), there stands the old farmhouse of Craigybwldan. The name is recorded from the sixteenth century onwards: *Grayg boulden* 1583, *Graige y bwlden* 1650, *Craig y bulldan* 1724, *Graig y Bulden* 1754, *Graig y Buldan* 1764, 1799, *Graig y Bwlden* 1803, and it is the second element which has caused speculation among enquirers.

The first element, W *craig*, is a common enough place-name element though it should be borne in mind that it is not always applied to a 'rock, boulder, stone' as a specific feature or 'steep rock, precipice', and that it can refer to a location where an outcrop of rock or stone is a feature, This could have been the case here as there are old quarry and colliery sites nearby.

Where the second element, *bwldan*, is concerned, I was informed by a correspondent who was brought up in Waunarlwydd that the belief locally was that the second element of the form *bwldan* (whatever may be said of *bwl-*) was W *tân* 'fire' and that in conjunction with *craig* the whole referred to a 'beacon hill'. However, the truth, as always, is more mundane as is indicated by a fortunate entry in the will of Philip Mansel of Llanddewi in Gower, 1553, which records the fact that Nicholas Balden owed the deceased the sum of 20s. It is, therefore, originally an English surname and may well be testimony to the fact that a family of that name were of that locality although nothing further about them is known to the present writer.

Further, it would appear that Craigybwldan was an alternative name for *Alte buldan*, which is also recorded in 1583 (= Allt Bwldan) and seems to occur more persistently than Craigybwldan in sixteenth-century sources: *Alkt Boulden* 1585, *Alt Buldan* 1590, *Alt Bulden* 1594. The 1585 form illustrates a non-Welsh scribe's difficulty with the W fricative *-ll-*, still evidenced in the unsuccessful efforts of speakers on radio and television to pronounce W place-names in which it occurs. W *allt* can mean 'steep, rough slope, cliff' as well as its more normal meaning in south Wales of 'a wooded slope' ('hill, ascent' in the north), and the 1884 six-inch Ordnance Survey map shows quite clearly the wooded nature of these slopes before they were cleared.

Another feature illustrated by this name and which also seems to confirm the interpretation favoured here is the use of the W def.art. before a personal name or surname which either is or appears to be of non-Welsh origin when it is included in a place-name. In Glamorgan, there are names like Coed y Cradock, Llancarfan (regarded as an English name despite its origin, Caradog), Tir y Barnard, Llantrisant, Tir y Byrbais 'Burbage's land' now Bryn Byrbais, Llanilltud Faerdre, Tir y Bwhayen (Bohun), St Fagans etc. Dr B.G. Charles notes several examples in Pembs., including Parc-y-Prat, Coed y Devonald (like Cradock, a surname which is a form of an original Dyfnwal), Pant y Philip etc.

Croes Cwrlwys : Culverhouse Cross

As Croes Cwrlwys is the form of the name adopted by a television company as the Welsh name of its headquarters on the western fringes of Cardiff, and since I bear some small measure of the responsibility for drawing attention to the form in the first place, it would seem incumbent upon me to provide an explanation of its provenance.

The English equivalent, of course, is Culverhouse Cross, the cross referring to the point where the road from St Fagans to Wenvoe crossed the portway from Cardiff to Cowbridge rather than to a wayside or boundary cross, as its earlier forms imply, *The Crossway* at the end of the seventeenth century and *Croesheol* 1762 (the W adj. *croes* 'cross' + *heol* 'way, street, road'). This was the forerunner of the complicated modern road junction. By 1885 it was referred to as Culverhouse Cross, incorporating the name of the adjacent farm, *Culverhowse* 1533, *Culver house* 1631, 1761, *Culverhouse* 1654–5, *Culferhouse* 1813–14, which stood in the detached portion of the parish of St Fagans a little to the south of the present Glan Ely School, its site now obscured by suburban housing.

The first element in the name of the farm derives from OE *culfre* 'a dove', the ME *culverhouse* 'a dove-cot, columbarium' being a well-attested place-name in England. Its literal Welsh equivalent is *colomendy* (< *colomen* 'dove' + *tŷ* 'house') but this was not used as the basis of the Welsh form of the name in this instance. By the eighteenth century it is clear that a colloquial form had evolved which was a Welsh oral rendering of the English form rather than a borrowing, namely *cwrlwys*. An estate plan of 1776 shows *Cwrlwys*, and a lease of 1786 refers to fields called *Pedwar Erwr Cwrlwys* and *Dwy Erwr Cwrlwys* 'the Cwrlwys four acres' and 'three acres' respectively. It developed by loss of the unaccented second syllable of *culverhouse* and a metathesis of the consonants *l-r* to *r-l* together with what is the normal adaptation of E *house*, OE *hūs*, in Welsh borrowings where it occurs as a second element in a compound, namely *-(h)ws*, as in *wyrcws* < E *workhouse*, *becws* < E *bakehouse*, *warws* < E *warehouse* etc. to give the form *cwrlws* and then, in this case, *cwrlwys*. The form *Curlass*, used by Emanuel Bowen on his map of Glamorgan in 1729, is malformed.

The diphthongisation of the final *-w-* to *-wy-* has been explained as a conscious over-correction based on the known colloquial reduction of diphthongs to the value of the dominant vowel component in some words, as in W *troed* > *trōd*, W *maes* > *mās* etc. (see p. 201). For instance, W *eglwys* 'church' is almost always pronounced *eglws* in the vernacular, so that *cwrlws* could be judged to have been the spoken form of an original *cwrlwys*, this having been erroneously 'restored'. Furthermore, it has been argued also that *cwrlws* is to be regarded as a common noun in this compound and should have the W def.art. before it to make it specific as the distinguishing element, thus *Croesycwrlws*.

Such contentions verge on the hypothetical. The most obvious influence on the form which place-names assume over the years is oral pronunciation by generations of people unversed in philological or phonological principles. Usage is the determining factor. The non-Welsh scribes and surveyors who produced estate maps, surveys and rentals were not likely to be sufficiently aware of the subtleties of Welsh sound changes as to be able to correct or over-correct forms of the place-names which they wrote. There is any amount of evidence to show that they almost invariably wrote what they *heard*, as best they could, in this case *Croescwrlwys*. It may be significant, for instance, that in the case of Tarrws, in the not-too-distant parish of Wenvoe, which is a rendering of E *tarehouse* (having *tare* 'vetch' as its first element to signify a place where cultivated vetch was stored for animal fodder) the earliest form recorded is *Tarwis* 1416–17 in a minister's account of the Glamorgan estate of the de Raleigh family in the Somerset Record Office. Further, Professor Melville Richards has testified to the origin of the form Peulwys, the name of a farm in the parish of Llysfaen, Denb. It is not based on an English form in *–house* but is an abbreviated form of an earlier Llys Mab Pilws 'the court of the son of Pilws', the personal name Pilws having become Peulwys in the vernacular.

If doubt is to be cast on the structure of the form Croes Cwrlwys, it may as well be said that the Welsh form of the name of Cardiff, Caerdydd, has no standing because it is not Caerdyf, the correct form, or Bargoed, which should be Bargod, or Corwen, Merioneth, which should be Corfaen, and a host of others whose present forms have been determined as the result of colloquial sound changes. Naturally, it is more convincing if such changes can be attested in recorded forms, and it may be granted that such forms as exist of *Croes Cwrlwys* are few, but no recorded form of *Croesycwrlws* has been found as yet. In addition, if the Welsh rendering of Culverhouse Cross is of comparatively recent creation, let us say no earlier than the middle of the nineteenth century, it could well be a simple 'unlearned' direct rendering of what the English form appears to be, without knowledge of the true significance of the distinguishing element, thus accounting for the omission of the Welsh def.art. before that element. Croes Cwrlwys, therefore, has attested usage in its favour.

Croescadarn

In the wake of suburban expansion in the parish of Llanedern on the eastern limits of Cardiff the naming of some newly created streets cannot be regarded as being inspired or appropriate. One outstanding example, however, is of considerable interest as a survival of a much earlier period although meaningless in its present changed form.

This road runs from the vicinity of the BUPA hospital northwards to cross a well-used country road which runs from Lisvane to St

Mellons. Its present name is Croescadarn Road, and several estate documents of the Kemeys-Tynte family of Cefnmabli confirm that this crossing (W *croes* 'cross') accounts for the first element of the name.

The modern form may give the impression that the second element in the name is the Welsh adjective *cadarn* 'strong, powerful', but *croes* is a feminine noun and as such demands that a following adjective which begins with a mutable consonant should be lenited initially to give *croesgadarn* which obviously does not occur in this case.

The explanation for this is made clear in the earlier recorded forms of the name. These are *Croesychadam* 1685–6, *Croes ych Adam* 1698, 1711–12, *Croes ach Adam* 1713, and so on consistently through the eighteenth century with occasional variation in an attempt to derive meaning, like *Croesuwchadam* 1781 and the like.

Two elements clearly follow *croes*, the second of these being the personal name Adam, in that form and not its Welsh equivalent Adda. Further, it is also well known that *ab* or *ap* (colloquial forms of W *fab*, the initially lenited form of *mab* 'son') was used with Welsh personal names to show the relationship of father and son: Dafydd ap Gwilym, Llywelyn ab Iorwerth etc. Similarly W *ferch*, the lenited form of *merch* 'daughter' was used in the same way to show that relationship. In everyday speech, the initial *f-* in *ferch* was also lost, and as GPC makes clear, several variations of the remaining *-erch* are attested in documentation and literary sources, such as *ych*, *ach*, *arch*, *ech* and *uch*.

The first two of these are present in the forms quoted above so that the implication is strong enough to be accepted that the original name of the road crossing in this particular case was *Croes ferch Adam* 'the cross(roads) of Adam's daughter'. We know nothing of Adam, nor the name of his daughter, but some corroboration of his existence is found in a field name in the locality, *Gwaun Adam* 1650 'Adam's moor'.

In their standard volume on *Welsh Surnames* T.J. and Prys Morgan suggest that the final *-ch* of *(f)erch* and its variants could possibly attach itself to the Christian name that follows to form a surname in the same way as the masculine *ab* gave Bowen from *ab Owen*, or Prichard from *ap Richard*. Here, possibly, *ych* (*ach*) Adam could have produced a colloquial form like *-chadam*, pronounced by non-Welsh speakers with a 'hard' *c* > *-cadam*. It would only be a short step from

Croescadam to the present Croescadarn, possibly aided by scribal error, *-am* being read as *-arn*.

Crofftygenau

This is at present the name of a substantial dwelling by the side of the road leading from the site of the old Rhydlafar Orthopaedic Hospital (now overbuilt) to St Fagans which itself bears the name of Crofftygenau Road. Earlier, it was the name of the farmstead, certainly in existence in 1570, on or near the site of the present house. This is open ground, sloping gradually downwards towards the south which makes it difficult to envisage the significance of the existing form of the name where W *genau* 'mouth, opening' seems to occur as its second element, even in the elliptical sense of 'an opening to a valley' or 'a break in hilly surroundings'.

The first element, W *crofft*, is a borrowing as a female noun of E *croft* 'a small enclosed field' arable or pasture, and has developed its own plural forms, *crofftau* or *crofftydd* which also appear in Welsh place-names, particularly in the abbreviated oral form *croffta*. The further treatment of the word in Welsh is also interesting because its initially lenited form after the Welsh def.art., *grofft* (< *y grofft*, as is normal with Welsh feminine nouns) was erroneously taken to be its radical form and was sometimes itself further lenited to give the form *y rofft* in a number of names in north Wales, and even *yr offt* by false division, with the full form of the def.art.

The form of the name in a 1570 survey is *crofte y ginyn*, and by 1720–6 *Croft y Ginni*, with *Croft y Gunny* 1728 etc., where loss of final *-n* in the second element is quite noticeable with a subsequent obvious uncertainty as to its exact nature. By 1834 it had become *Croft y guinea*, but fortunately intervening seventeenth-century forms in the Bute estate papers help to solve the problem, those being *Krofft Eginin* early 17c., *Croft Eginin c*. 1670, 1671 etc. and on into the eighteenth century.

The second element is thus confirmed as W *eginyn*, with a strong accent on the second syllable, a diminutive form of the word *egin* 'young growth, shoot, sprout'. In the compound *crofft-eginyn* the first unaccented syllable of *egínyn* became indistinct in common speech and was probably mistaken for the vocalic form of the Welsh definite article *y*, thus accounting for the *crofte y ginyn* of 1570.

There is therefore no warrant for the present map-form. It was originally a descriptive name for a plot of fertile land, where young crops started into growth successfully, *Crofft-eginyn*.

Cwm Talwg

Various elements used with W *rhyd* 'ford' in place-names is a subject of comment elsewhere in this volume, among them colour adjectives to describe the yellowish-brown hue of the water of a well-used ford (see p. 166). Another is W *halog* 'dirty, muddy, discoloured' which is used also in Welsh river names, cognate with Ir *salach* 'dirty' in a name like Bellasallagh (*Beal-atha-salaigh* 'the entrance to the miry ford') and which can be compared with OE *fūl* 'foul' in Fulford, Fullabrook, Fullaford, Fullamoor etc. in Somerset and Devon. The W *rhydhalog* occurs frequently in north and south Wales, examples also occurring in the parish of Llangynwyd (*Nant Rhyd Halog* 1695–1709) and north of Ystradowen (*Ryd Halocke* 1628–9, *Rheed Talog* 1729, *Rytalog* 1750) in Glamorgan. In common usage, as shown in the second example, provection of the final -*d* of the first element is caused by the following initial *h*- of the second to produce the spoken form *Rhytalog* (cf. *Botalog* < *Bod* + *halog*, the name of an old township in the parish of Tywyn, Merioneth).

On the western fringes of Barry two streams flow down to Porthkerry Park, one of them the stream that gave the name to Cwmcidi, the valley down which it flows. This was the name which became that of a deserted village in the vicinity of the present Cwmcidi Farm, and of an extra-parochial church.

Although the name of the stream (W *nant*) was Cidi (John Leland specifically refers to *Kiddey mouth*, 1536–9) its name on modern maps is Nant Talwg ('Nant Talwg which runs through Cwm Cidy' is the Barry Centenary Book's description) and this has been the reason why its valley has acquired the name Cwm Talwg. However, the way in which the Talwg (= Cidi) came to be so named is based on a misconception and Cwm Talwg has now become the name of an urban area in Barry, with Nant Talwg Way as one of its streets.

The stream flows from its source in the contiguous parish of Merthyr Dyfan and is crossed today by Pontypridd Road, the B4266. It is highly likely that this road's predecessor was that which was called *Redhollocke Lane* or *Redhallock Waye* in 1622, a reasonable attempt by an English scribe to reproduce the W form *rhydhalog* as the first element. Neither is there much reason to doubt that *rhydhalog* here refers to the point at which the lane or way crossed the stream, a much used, miry and dirty ford. Maps of the Fonmon, Wenvoe and The Mount (Dinas Powys) estates confirm this nomenclature: *Reedhallock Lane* 1762, *The Reddhallock Way* 1763, *Rhyd Hay Lock* 1798, and forms in which provection of *d* to *t* occur are echoed in *Retalog Lane* 1778, *Retailock* 1783. In addition, an adjacent field through one end of which the stream ran is named *Rhyd Hay Lock* on the Wenvoe estate map of 1798, the stream being then shown running down into Cwmcidi.

It would seem therefore that what appears to be the form of the second element *talog*, with its colloquial variant form *talwg*, having the initial provection noted above, was mistakenly assumed to be the name of the stream rather than the derivative of the adjective which once described the nature of a ford on its upper reaches, thus begetting the forms Nant Talwg and Cwm Talwg at a later date.

Cwmbwrla

In place-names, W *cwm* 'valley, glen, coomb' is normally found in conjunction with a stream or river name and the *-bwrla* in this well-known Swansea example is no exception. Indeed, it was this realisation which probably led to its being known at one time by the name *Bwrlais*, this being a form deliberately contrived to make sense of the second element in the name by linking it to the W *glais* 'stream, rivulet' which occurs generally in Wales, but more particularly perhaps in the south. It is seen as a simplex form in the name of the village Y Glais (*Aber Gleys* in 1203, the location of the confluence of the stream with the river Tawe) or in combination with adjectives in specific river and stream names like Morlais, Gwynlais, Marlais, Dowlais, Dula(i)s etc.

The stream which runs through Cwmbwrla turns southwards to run into the Tawe at a spot which was not far from the Cambrian Pottery of former times. Also, in the early years of the existence of the borough of Swansea the stream was its northern boundary and is named in the first charter granted to the burgesses sometime between 1153 and 1184 as *Burlakesbrok*, and in a later charter *Burlakysbrok* 1306. The final element of this form is OE *brōc*, the modern *brook*.

However, there was no real need to add *brook* to *Burlake* because the second element *-lake* here is not what is meant by *lake* in ModE (which came into ME from Fr *lac*, Lat *lacus*) but rather that which is seen in English names like Mortlake, Fishlake, Shiplake etc. < OE *lacu*, which Margaret Gelling defines as a 'small, slow-moving stream, side-channel'.

Further, as noted above, this stream was part of the borough boundary and the OE form of *borough* was *burh*, having originally the sense of 'fortification' but later to include 'fortified town' and subsequently 'town, borough'. This is the prefix seen in the names of English towns like Burford, Burgate etc., so that *Burlak(e)* here can be derived from an OE *buhr-lacu* 'the boundary ditch, or stream' of the borough. In this case, the final 'hard' *-c* (*-k*) survived until at least the seventeenth century: *Purlocke brooke* 1584, *Burlax Brooke* 1685, *Burlocke's brooke* 1689, but that century also sees its gradual demise in common parlance for *Cwm Bwrla*, with its Welsh first element, had made its appearance by 1641 and 1650.

The name occurs in one other location in Glamorgan and is the name of a stream which partly encircled another old borough, Kenfig (Cynffig), which disappeared in time beneath the sand. Today, the name of a bridge (W *pont*) which carries the road from Margam to Mawdlam over that stream is Pont Bwrlac. In 1147–83 the stream was the *Bulluchesbruhe*.

Cwrt-y-fil

Cwrt-y-fil was a farmhouse on the southern fringes of the parish of Penarth and is today recalled in the street name Cwrt-y-vil Road, but what seems to be something of a mystery is the Cwrt-y-vil Castle shown as standing behind the farmhouse on the old six-inch Ordnance Survey map.

The so-called castle cannot have been what is normally represented by the term (excluding its use in pompous 'folly' names of fairly recent vintage). A persistent tradition gives it a religious association. Samuel Lewis (1833) speaks of 'a ruin, now converted into a barn, which was formerly a chantry chapel', and David Watkin Jones (1874), influenced by the writings of Iolo Morganwg, refers to the existence in that location of an ancient religious *côr* in the sense that Iolo used the term 'a religious community or convent'. In the Papal taxation of 1291 reference is also made to a chapel attached to Penarth.

It is known that the manor of Penarth came into the possession of the abbey of St Augustine's, Bristol, before 1183 and by 1540 there is recorded the existence therein of *Canon Court* which was the *scite of the Mannour and Barton*. The estate papers of the earls of Plymouth confirm the existence of the *Barton called Cannon Court* in 1745–52. The use of *canon* in the name appears to confirm the religious association and certainly *barton*, which occurs frequently as a simplex place-name in England, OE *bere-tūn*, basically a 'barley enclosure' or farm, but its most usual sense in ME was 'an outlying grange' for storing the lord's crop hence often simply 'a grange, a demesne farm', cf. OE *bere-wīc*, with *wīc* in Barwick and the common Berwick.

In south Wales (whatever may be said of the north) one of the ways of naming monastic granges is the reasonably frequent use of W *cwrt* 'court' in their Welsh names (another word used being *mynachdy*), most notably the names of the several granges of the abbey of Neath—Cwrtrhydhir, Cwrt-sart, Cwrtycarnau, Cwrtybetws—and it is suggested here that this is the significance of *cwrt* in Cwrt-y-fil, a 'grange' or *barton*.

The second element is not so easily disposed of. Unfortunately, no forms of the name as it stands earlier than *c*. 1700 have been attested. That is the date of the forms *Cwrtyvill*, *Court-y-vill*, then *Court a Vill* 1787, and, possibly more important, *Courtfield* 1789, 1793, *Courtefield* 1790. The form used by Iolo Morganwg was *Courtville* 1799, with no attempt to make a Welsh name of it and, as Professor G.J. Williams consistently emphasised, when Iolo made no attempt to interfere with the form of a name or to cymricise it as something of his own invention, it can be taken as reasonably certain that he would use the form current in common usage in his day.

As noted elsewhere (p. 212) several names in the Vale of Glamorgan which end in *-field* have been and are pronounced as if that element were *-ville* under the influence of the spoken forms of the south-west of England where initial *f-* is often voiced to *v-*. It is possible, therefore, that a form like *Courtfield* could have been pronounced locally as *Courtville*. Further, another feature of south-western English dialectal forms is the appearance of an intrusive and unaccented medial *-a-* between the constituent elements of a name. This seems to be echoed in the 1787 form *Court a Vill* and to judge from the forms attested subsequently what seems to have happened is that in common usage the intrusive vowel was mistaken for the vocalic form of the W def.art. to produce *Cwrtyfel* 1813–14, *Courty Vill Farm* 1819, *Courtyfil* 1826, *Court y Phil*, *Court y Fill* 1829 formalised as *Cwrt-y-fil* on the 1833 first edition of the one-inch Ordnance Survey map.

This interpretation is to be preferred to that which regards the second element of the name as the W noun *mil* 'animal, beast, creature' which is not current but is to be seen as an element in compounds such as *bwystfil* 'beast', *morfil* 'whale', *milgi* 'greyhound' etc. (not the numeral *mil* 'one thousand', see p. 153). This is further supported by the fact that *mil* in this sense is a masculine noun and would not be lenited initially after the W def.art. which, in turn, confirms that the development of the form would appear to be entirely phonological.

Cwrtyrala

The map form, *Courtyrala*, is the anglicised version of the name of this mansion in the parish of Michaelston le pit (Llanfihangel-y-pwll), Cwrtyrala being invariably the form used by the local inhabitants in common speech.

Several attempts have been made to interpret the name over the years, including a suggestion made by the renowned Edward Williams (Iolo Morganwg). He uses the form *Court yr Alaw* in 1796, but it is not clear what he meant to imply by this, whether he intended the second element to be understood as W *alaw* 'water lily', or whether he was responding to that musical whim which caused his word-peddling contemporary, William Owen-Pughe, to ascribe to the word the meaning of 'air, music, tune' of his own invention. In either case, he was in error.

Further, earlier in the eighteenth century an effort had been made to adopt the Welsh word *elerch* as the second element of the name in the forms *Cwrt yr Elerch* 1784, *Cort yr Elerch* 1786, in the mistaken belief that it was perhaps a variant form of *eleirch* or *elyrch*, the accepted plural forms of W *alarch* 'swan'. By the end of the century the form *Cortyralla* 1792, *Courtyrala* 1818, *Court yr Alla* 1820 etc. had become the norm but still with some uncertainty over the meaning of the second element of the name, with a strong leaning, possibly, towards the full form of the Welsh definite article *yr* + a borrowing into Welsh from the Middle English *alley*, namely *alai*, in its colloquial form *ala*, as in names like Alafowlia 'bowling alley' (Denbigh), Yr Ala (Pwllheli), Pen-yr-ala (Tregarth) and Caernarfon's Ala Las.

This interpretation, however, is disproved by the earlier forms of the name echoed in one late form used in 1844, *Court y Rala*, the point being that what we have here is not a form that contains the full form of the definite article *yr* + the element *ala* but, rather, one that has the vocalic form of the article *y* + an element which begins with *r-*, *y rala*. The earliest forms attested are: *Court Rayle* 1578, *Court y Raylley* 1657, *Courtralley* 1670, *Court Raille* 1674 etc. which confirm the connection with the property of the family who bore the locational surname of Rale(i)gh, Rawley, from near Barnstaple, Devon (of which the well-known Elizabethan Sir Walter Ralegh was a member) into whose hands the manor of Michaelston le pit fell at the end of the thirteenth century.

The name as it stands, Cwrtyrala, is wholly Welsh, this being made obvious by its syntactical structure. Were it an English name, it would be a form similar to *Raylescourte* which is recorded in 1283, but did not survive.

Y Cymin : The Kymin

One of the names which became the subject of speculation as to its meaning when publicity was given to the refurbishing of Penarth pier was that which appears generally on maps and locally as The Kymin, once the name of a smallholding on the left bank of a small stream where its outlet into the Bristol Channel forms a gap in the cliff a little to the north of the pier and below the present Beach Road.

This situation led one observer to postulate a Scandinavian origin to the name, claiming that it contains ON *mynni* 'the mouth of a river where it runs to the sea'. Earlier forms are: *Kimming Farm c.* 1700, *The Kemmin* 1730–1, *Kiming* 1799, *Kymin Cottage* 1808, *Cymmyn* 1813–14, *Kymin* 1844, none of which are of sufficiently early provenance to support this view.

Indeed, there is evidence in 1730 that the name was originally that of the land on which the farm stood, thereafter used, as in so many cases, as the name of the farm itself. It is described as 'the low land sloping to the shore at the beach west of Penarth Head', the reference to the location in some documents being specifically to common land. This would seem to identify what is referred to in sixteenth-century documentation relating to the land of St Augustine's Abbey, Bristol (in whose possession *terra de Pennard* had been since the end of the twelfth century, see p. 59), namely *Le Comon* 1540–1, *the Comen* 1542.

It can be accepted that the name is a variant form of the W borrowing *cwmin, comin* from ME *cumin, comin* 'unenclosed land held in common by the inhabitants of a community' (indirectly from Lat *commune*, probably through an OFr form), this being *cymin*, in English orthography *kymin*. It may well be that earlier references occur in the personal name *Willelm de Keymin* 1238 in a document relating to the nearby church of Cogan, and the *villa de Gymyn* in 1429, *in comitatu de Cardyff*. It should be noted, however, that the W form *cymin* is not listed in GPC, although T.C. Evans (Cadrawd) (1846–1918), the noted antiquary and folklorist, was firmly of the view that *cymin* (*kymin*) was an authentic Gwentian form corresponding to the synonymous *cimla, cimdda, cymdda*, in Glamorgan (see p. 42).

This emphasises the relevance of the name of the prominent hill which overlooks the town of Monmouth in Gwent, also The Kymin, which Archdeacon William Coxe in his *Historical tour through Monmouthshire* (1801) asserts was unenclosed common land. This is infinitely more acceptable than Sir Joseph Bradney's suggestion that it is a form of the W *Cae-main* 'stony field'.

Cynffig : Kenfig

Although Kenfig is not an entirely satisfactory modern representation of the correct Welsh form Cynffig, as a written form it has a long pedigree. Its predecessors, forms like *Kenefec, Kenefech, Kenfeck, Kenefeg* etc., which appear from the twelfth century onwards are probably genuine attempts by generations of non-Welsh scribes to render the sound of the name as they heard it spoken. Before such forms are denounced as corruptions, as they so often are, it must be borne in mind that in days when there was no orthographic standardisation the reproduction of unfamiliar names—unfamiliar, that is, to the ear of the scribe—was an oral process. The W fricative *–ff-*

was represented by E *-f-*, and the W prefix *cen-* had a colloquial variant form *cyn-*, as in a pair like Cencoed and Cyncoed.

In Cencoed, *cen-* is a colloquial shortening of W *cefn* 'ridge, back' in compounds, the original form of Cencoed, and its variant Cyncoed, is Cefn-coed, with which can be compared *cenffordd* for *cefnffordd* 'highway', *cenros* for *cefn-rhos* 'ridge moor' (which became Gendros, see pp. 82–3), and Cendon for Cefn(y)don near Tonna, Neath. Because of this, there is a tendency to see a similar process at work in the form Cenffig (Kenfig) or Cynffig, and to interpret the first syllable here also as W *cefn*, but this would be an error.

There can be little doubt that Cynffig was originally the name of a river which rises on Mynydd Margam, the *aqua de Kenfeg* of the early documents and John Leland's *Kenfike Ryver* 1536–9, hence the related Abercynffig (*Aberkenfigg* 1596–1600) with W *aber* 'confluence', Blaencynffig (*Blain Kenefeg* 13c.) with W *blaen* 'source', and Cwm Cynffig (*Come Kenfyg* 1527). This was the river which ran into the Severn Sea past the small borough which took its name but was for centuries inundated by drifting sand. This naming process is reminiscent of the Roman way of naming towns after the rivers on which they stood, like Neath, *Nidum*, and Loughor, *Leucarum*.

What gives Cynffig its peculiarity, however, is that in all probability it is basically a personal name. As R.J. Thomas showed clearly in his study of Welsh river names (1938), any number of Welsh river or stream names, the latter in particular, have their origin as personal names, some of them of such antiquity as to be no longer in circulation. This is true of Cynffig, but we do have record of it as the name of a witness in the *Liber Landavensis*, in the OW form *Conficc*, *Cinfic(c)*.

The prefix *cyn-*, *cen-* in this name is therefore not W *cefn* but an integral part of an old Welsh personal name as found in other personal names like Cynfal (Cynfael), Cynwrig (Cynfrig) and Cynan, all of which, as it happens, are also stream names in Merioneth, Brecs. and Glamorgan.

Cyntwell

This is the present form of the name of an urban area in the old parish of Caerau on the western fringe of Cardiff. It has an interesting etymology which cannot readily be deduced from its modern form, based as it is on the form *Saintwell*, the original *S-* having been replaced by the sibilant *C-*. It is the form *Saintwell* which appears on the six-inch Ordnance Survey map of 1884, but on the first edition of the one-inch map of 1833 the form is, significantly, *Saintwall*.

The vowel variation, with *-wall* becoming *-well*, the result of loose pronunciation of an unaccented syllable in common parlance, is not difficult to appreciate (and it is to be noted that such is the order of change, not *-well* to *-wall*) particularly as the popular view might the more readily associate saint with *well* rather than *wall*.

In the parish of Llanrhidian, in Gower, a farm called Llodrog (Llotrog) near the church of Llanynewyr (see p. 117) appears as *Llodrog alias St Wall* in a survey of 1764, and by 1884 the alternative name is written in its proper form *Saintwall*, but it is now reasonably certain that this is what is represented in an earlier survey of 1641 by the W form *Seintwar*. This form is a borrowing, with loss of the final syllable, from a ME form like *seintuarie*, now *sanctuary*. In common speech there is a tendency for the final *-r* of an English disyllabic word to become *-l* in Welsh borrowings as in *corner* > W *cornel*, *dresser* > W *dresal*, *razor* > W *rasal* etc. This happened in the case of *seintuar(ie)*, *seintwar* to produce *seintwal* which was in turn anglicised and understood as *saintwall*.

The meaning of *sanctuary/seintwar*, of course, apart from its basic sense of 'holy place', is 'a church or other sacred place in which by the law of the medieval church a fugitive from justice was immune from arrest'. This right was also claimed by the Knights Hospitallers of the Order of St John. At one time the rector of Llanrhidian was perceptor of the Order at Slebech, Pembs., and this would seem to help to explain the existence of a sanctuary at Llodrog near Llanynewyr, also an old foundation in the parish.

The location of the sanctuary in the parish of Caerau was near Culverhouse Cross, not far from the site of the now demolished farm of Sweldon on rising ground well to the west of the old Caerau parish church. The farm site is indicated by the present street names Farmhouse Way, Upper Meadow and Lower Acre. Further down the hill is a looped road called The Sanctuary, with Sanctuary Court leading off it, which seems to indicate approximately the site in question, for this looks like the location of the building named *Saintwall* on the 1833 one-inch map, modern housing and commercial premises having obscured most of the earlier features of the landscape. It is called a cottage in 1695 and 1787 and many references occur to it in the diary of William Thomas of Michaelston super Ely (1762–95) as a property quite distinct from the neighbouring Sweldon. In 1767 the diary contains a reference, 'the house of Rachel Elbred, widow, of the Sanctuary, was broken, and most of her apparel stolen … '. Further, it is also known that this property was a detached portion of the manor of Milton, Pen-coed near Bridgend, once in the possession of the Hospitallers of the Order of St John.

Here, therefore, is another *Sanctuary*, W *Seintwar*, which became *Saintwall*, then *Saintwell* before the creation of the dispensable modern Cyntwell.

Drope

When Aneirin Talfan Davies wrote the first Welsh account of his wanderings in the Vale of Glamorgan (1972) he drew attention to what he called the 'strange' name of this village in the parish of St Georges, near Cardiff.

The name, as recorded in 1535, was *Britton super Elly* and *Brytton upon Eley* in 1578. In 1540 there is a reference simply to *Le Thorpe* and *Thorpe* further confirmed by *Britton-Thorpe* 1657–60, *Brytenthorp* 1674–5.

The precise significance of *Britton* here remains uncertain. It could derive from OE *brycg* 'bridge' + the common suffix *-ton*, OE *tūn* 'farm, farmstead', to signify the farm by the bridge on the river Ely but which can hardly apply to the present location of the hamlet of Drope. It could also have, on the other hand, a connection with the coterminous parish of St Brides super Ely north of the river; indeed, some commentators equate Britton super Ely with the village of St Brides. There occurs also an unidentified *Brideton c.* 1291 and an *ecclesia de Brideston* in the *Liber Landavensis* (which has been thought to refer to St Brides Minor, north of Bridgend) which could be the basis of the later form *Brit(t)on*.

Be that as it may, it is the presence of the element *thorp* which is relevant to the meaning of Drope. Here, it is probably the ME *thorpe* 'outer farmstead, hamlet', its provenance being later than what would be suggested by its cognate Old Norse form *Þorp*. Frequently, in common usage, the metathesised form *throp(e)* develops as an alternative, and it is such a form which has seen the substitution of *dr-* initially for *thr-* to produce the form Drope, more specifically, later The Drope.

The significant point to be made in this connection is that such a substitution is a feature of the dialects of the south-west of England in particular but including the wider range of the counties of Gloucester, Worcestershire, Dorset, Bedford and Hampshire. With Drope can be compared names like Drupe (Devon), Pindrup (Gloucs.), Droop (Dorset) and the like. It confirms the considerable influx of elements of the population of these English regions into the area, especially into the environs of Cardiff, the Vale of Glamorgan and Gower, their dialects having been a strong influence on the phonology of the toponymy of these localities over the years.

In the case of Drope one further development can be discerned. By the eighteenth century the form bears the indications of a strong Welsh influence on its form. To correspond to the more specific English The Drope there appeared the Welsh *Y Drope*, or *Y Drôp* in Welsh orthography, together with a gradual change of gender shown by mutation of the initial consonant after the Welsh definite article, *Y Ddrôp*, and further, the appearance of an epenthetic vowel which is a feature of the south Wales vernacular, as with W *ofn* 'fear' > *ofon*, *dwfn* 'deep' > *dwfwn* etc. to give the recorded form *Y Ddorôp*, with the accent on the final syllable.

Drysgol

The normal meaning of the W adj. *trwsgl* is 'awkward, clumsy: bungling' but as Sir Ifor Williams showed years ago, although it acquired various shades of meaning, it is in the sense 'rough, rugged' that it is used as a descriptive element when it occurs in the names of hilly or mountainous locations. The feminine form is *trosgl*, and to give it a substantive sense as the name of a particular hill or mountain, preceded by the W def.art. it is lenited initially to give *Y Drosgl*. This occurs in a number of instances, the best known perhaps being the name of a prominent height near Llanllechid, Caerns. (contrasted, significantly, with the nearby Y Llefn, the feminine form of the W adj. *llyfn* 'smooth'), and in another location south-east of Llanfairfechan, Caerns., near Bwlch y Ddeufaen. In common speech it becomes *Y Drosgol*, having acquired an intrusive vowel to form a second syllable, and occurs in that form near Dolwyddelan, Caerns., and in the Creunant Forest on the northern shoulder of Mynydd Marchywel, south-east of Ystalyfera. *Y Drysgol* could well be a variant form. Sir Ifor notes Y Drysgol Goch near Cynwyl Elfed, Carms., to which may be added Drysgol above Glanaman, Carms., between Rhaeadr and Llangurig, Radnor, beside the Nant-y-moch Reservoir and near Ystrad Meurig, both in Cards., and with a further intrusive vowel in the form Dyrysgol near the Aran Fawddwy south west of Llanuwchllyn, Merioneth, these examples being merely a selection of what can be evidenced.

The name cited here, however, has a different etymology and is that of a holding in the parish of Radur, near Cardiff, later divided into two farmsteads (the large and the small, W *mawr* and *bach*) but which have both disappeared by the present day leaving a trace in the name of Drysgol Road.

In 1766 it occurs in documentation as *Dyrys-coed*, with a further reference to *Dyryscoed Fields*, then *tryscol* 1783 in the land tax assessments, followed by *Drusgol, Drwsgol* 1784, *Druscol* 1786, *Driscol* 1787, *Dryscol, Drysgol vawr* 1808 etc., the first edition one-inch Ordnance Survey map excelling itself by showing *Yrysgol* 1833 (as if it were the W def.art. *yr* + *ysgol* 'school'). This name is compounded of the W adj. *drys* 'wild, rough etc' which became *dyrys* in common usage, again with an intrusive vowel, and W *coed* 'wood, trees, forest' etc. to give the combined meaning of 'a thick-branched, dense, thickset plantation, or grove' very similar to another fairly common name, Dryslwyn or Llwyndyrys with W *llwyn* 'grove, copse'. In this particular case, the diphthong *-oe-* of the second element has been reduced to *-o-* (a feature of the local dialect (see p. 201). More exceptional is the additional change of final *-d* to *-l*, but the collected evidence shows conclusively that this is, indeed, what has occurred in this instance.

Not every example of a name which appears as *drysgol*, or contains that form as an element, therefore, can be interpreted in the same fashion, thus emphasising the need for a close examination of the evidence in each case. One further interesting example occurs in the neighbouring parish of Llanilltud Faerdref (Llantwit Fardre), namely the name of the farm Drysgoed. It is the name of an original holding of land on the northern slopes of Mynydd y Garth, *Tyre y trysgoed* 1558 (with W *tir* 'land') or *Tyre keven tryscoed* 1567, *Tyr Trescoed* 1570, which was also subdivided by the eighteenth century into *Driscol Vawr* and *Driscol Vach* 1725, the forms used in local speech, although the form *Drysgoed* continued to survive in documentation, being *Driscoed* 1725, *Tir y Dryscoed* 1728, etc. until the present day in the case of the former of the two properties. It was strongly emphasised to me by a friend whose wife was the niece of the farmer who held that property in recent times that the name was always pronounced *Drysgol* by the family.

Meanwhile the name of the other portion, the *Drysgol Fach*, was already showing signs of change to *Pryscoed* in the eighteenth century, being *Tir y Priscoed* 1709, 1725, *Pryscoed* 1791, *Prys Coed* 1844 and *Presgoed* on the Ordnance Survey map, with W *prys(g)* 'copse, grove, plantation' introduced as first element, probably in order to achieve a recognisable meaning with *coed*. The tithe award of 1844 lists the two farms as *Dryscoed* and *Pryscoed*.

Also recorded in 1809–36 is the form *Tyrysgol* for *Drysgoed* which is reminiscent of the Radur *Yrysgol* 1833, but it may be worth considering that, apart from the possibility that an analogy with W *ysgol* 'school' may have contributed to its appearance, it may be only a scribal error based on the variant form *Dyrysgol* noted above.

Ewenni

One river name in Glamorgan which was not adopted as the name of a Roman fort on its banks and is recorded in one of the early classical sources is that of the Ewenni which runs down past Pencoed, Llangrallo (Coychurch), the priory of Ewenni and into the Ogwr near the joint estuary at Aberogwr below Merthyr Mawr. The source is the *Cosmography* of the anonymous monk of Ravenna *c*. AD 700 and the form of the name is *Aventio* which, like many others of its kind, has received considerable attention from distinguished scholars. It is possibly the oblique case of a Romano-British *Aventius* which survives as Ewenni by ultimate i-affection over two syllables (*Euenhi* in the *Liber Landavensis*) with a variant form *Ewennydd* (Giraldus Cambrensis in his *Itinerary in Wales* (1191), has the form *Ewennith*). A series of continental river names like Avenza (Italy) and Avance (four in France and one in Switzerland) suggest a possible feminine derivative, *Aventia*, as their origin. Sir Ifor Williams rejected this form for Ewenni on the grounds that such a form would give **Eweint* or **Awennedd*. The W form *Ewenni* demands a British terminal **-ios* (masc.) or **-ion* (neuter) and since the Ravenna *Aventio* is an oblique form of *Aventius*, it probably represents a Br **Aventios* which diminishes the problem of form. Professor Kenneth Jackson went a step further back to suggest that the suffix could have been **–īsos*, **–īson*.

As yet, there is no unanimity about the meaning of the name. It has been pointed out that the form *Aventia* could be a pers.n., possibly that of a female deity, but whether *Aventios* has some similar connotation is not established. One suggestion is that the female form *Aventia* has a root-form meaning 'water, spring', another that it is that which is seen in Lat *avus* 'grandfather' in its feminine guise and that *Aventia* suggests 'little grandmother' in the same way as *Matrona* can be seen in the name of the river Marne in France. Yet another favours the existence of a root-form seen in Lat *aveo* 'to wish, desire' and W *ewyllys* 'will'.

Later recorded forms of the name are *Ewenny*, *Ewni* 1149–83, *Eugenni* late 12c., *Eywenny* 1291, *Ewenny* 1348 etc. to *Wenny River* 1536–9 and *Y Weni* 1606 etc. these two latter forms showing the development of the name in the vernacular with an erroneous substitution of the W def.art. *y* for the unaccented first syllable or indeed its omission, cf. *Aradur* > *Y Radur* > *Radur*, *Radyr* (pp. 159–60).

Y Fanhalog

The form of the name of this farm in the parish of Llanwynno on the Ordnance Survey maps is Fanhaulog, where Fanheulog would be more accurate, based as it is on the assumption that the name is compounded of W *man* 'place, location' (lenited initially after a missing W def.art.) + the W adj. *heulog* 'sunny'. This assumption was frequently made in nineteenth-century documentation, *Fan-heulog farm* 1826, 1832, *Fanhaulog* 1850 etc., but it has no validity.

The correct form is Y Fanhalog, a variant form of *Y Fanhadlog*, the basis of which is W *banadl* 'broom' + the collective ending *-og*, which when added to the names of plants and vegetation, as in *rhedynog* (W *rhedyn* 'fern, bracken'), *eithinog* (W *eithin* 'gorse, furze') etc., signifies a place where such growth is plentiful. The strong accent on the penult being marked by the growth of *-h-* before the vowel, *banadl + og > banhadlog*.

A further colloquial change is evident in the variant form Y Fanhalog, based on the form *banal* for *banadl* where *-d-* is lost in the combination *-dl-* (possibly through an intermediate form in *-ddl-*) as in W *anal < anadl* 'breath', *cystal < cystadl* 'as good, equal'. There can be little doubt that this is the case as confirmed by the forms *Vanhaddloge* 1633, *Vanhalog* 1783, 1787, 1791–1826, *Vonallog* 1804 etc. and in his writings on the parish of Llanwynno (1878–88) William Thomas (*Glanffrwd*) invariably refers to *Y Fanhalog*.

Unfortunately, by the end of the nineteenth century the name began to be misunderstood as a form of Y Fynachlog (W *mynachlog* 'monastery'), being thus shown on the first edition of the OS one-inch map of 1833, a tendency which continues to this day and is perpetuated in the Glamorgan Inventory of the Royal Commission on Ancient and Historical Monuments in Wales.

A substantial parcel of land in the parish of Llanwynno was

granted to Pendar, an abortive daughter-house of the abbey of Margam about the middle or second half of the twelfth century but by the thirteenth century most of it had come into the possession of the abbey of Caerlleon or Llantarnam, Gwent, and across it, from east to west, ran an old pilgrimage route from Llantarnam to the shrine of Penrhys on the ridge between the two Rhondda valleys. It is probable that among the Penrhys complex of buildings the abbey had a grange farmhouse, and on the strength of the existence of another farm on the route in the parish of Llanwynno, not far from Y Fanhalog, which the Royal Commission cannot date any earlier than the mid seventeenth to eighteenth centuries, but which bears the name of Mynachdy 'grange' (literally W *mynach* 'monk' + *tŷ* 'house'), it can be accepted that this was the site of another similar establishment which could have dispensed hospitality on the route, or at least its approximate location. In documentation the earliest forms seen by the present writer are *Manachdy* 1562, *Manachdee* 1626, *Tire y Manachtee* 1652, *the Manachty* 1656–7 etc.

It is in this context that the misinterpretation of Y Fanhalog as Y Fynachlog has arisen, more specifically in connection with the identification of a chapel which is known to have existed in the area and named Capel y Fanhalog. This was wrongly assumed to be connected with Mynachdy. As noted above, the site of the farm of Y Fanhalog is in reasonably close proximity to Mynachdy. One recent claim identifies the site of the farm as that of the chapel and that its name appears 'in older documents' (unspecified) as Capel Fynachlog, adding the translation 'the monk's chapel' rather than 'the monastery chapel'. This is not the case, the chapel being located at National Grid reference ST 0511 9474 and the farm at ST 0481 9430. Furthermore, the Royal Commission's Inventory astonishingly informs us of 'a site called Capel Fanhalog, a corruption of Capel Fynachlog' whereas the reverse is true.

The source of such a belief is not entirely clear. David Watkin Jones (1874) refers to 'a house named y Fynachlog' (*tŷ o'r enw y Fynachlog*) and in his comments on the text of the grant of land and property to Pendar, Walter de Gray Birch in his *History of Margam Abbey* (1897) notes 'a site marked Fynachlog, or Monk's house south-west of Mynachdy'. The use of the word 'marked' implies a map source and sheet 36 of the first edition of the one-inch OS map, 1833, which has Fynachlog for Fanhalog, 'corrected' to Fan-heulog on the 1884 six-inch map, appears to be the culprit. In passing it is interesting to note that the person who advised the OS surveyors on the forms of place-names on the Glamorgan section of the one-inch map was the antiquary John Montgomery Traherne (1788–1860) of Castellau and Coed Riglan.

The reality is that no form of the chapel's name can be found in documentation earlier than the eighteenth century and that it bears no relationship whatever to the earlier monastic grange in the vicinity. In fact it refers to a Nonconformist meeting house, or 'chapel' in that sense. In 1786 a lease of a portion of the land of *Fenhallog Felan* was granted 'to build a Methodist meeting house for the Worship of God'. This was the Fanhalog farm which here has the element *felan* as part of its name, this being the colloquial female form of the adjective *melyn* 'yellow', that is, *melen*, which is not otherwise evidenced among the collected forms of the name but is obviously not an illogical addition to a name which signifies a place where there was an abundance of broom. This was the chapel which was registered and licensed in the bishop's court in 1811 as *Cappel y Van Hallog* and appears as *Capel Fan-heulog* 1832 and *Vanhayly Chaple* in the Religious Census of 1851. Such was the origin of the chapel known as Bethel, Llanwynno, built on the same portion of land as the old Capel y Fanhalog in 1876, which itself closed its doors to public worship in 1976. Both structures still stand converted into private houses.

One can only speculate as to whether Capel y Fanhalog could have become Capel y Fynachlog as a loose rendering in common parlance or as an error in documentation by non-Welsh scribes, the second element being mistaken for *-fynachlog* because of a superficial similarity of form, for in the Dunraven estate papers in 1778 the name of a well (W *ffynnon*) in the parish of Coychurch appears as *Funnon-y-munalog* for Ffynnon y Fynachlog (see p. 46).

Forty

The farm which bears this apparently misleading name stands beside the road from St Fagans to Peterston super Ely (Llanbedr-y-fro) and in close proximity to St Brides super Ely (Llansanffraid ar Elái). Close by stands the present *Willows Farm* which was *Forty Fach* 'the smaller, lesser Forty' in 1884, both these farms being portions of an original single holding. There is evidence that the form *Y Fforti* was in circulation at the beginning of the twentieth century, a local cymricised form of a basic English name, but the general tendency was to regard it as a shortened form of *forty acres*, presumably the assumed extent of the original holding.

No direct evidence has been found for the period of origin of the farm but it is reasonably certain that it is the property to which the truncated form *the Fort* refers in 1615 although admittedly in a source which cannot be regarded as wholly reliable. Subsequent recorded forms are *the ffortai* 1763, *Fortai* 1766, *Forty* and *Lower Forty* 1784 (probably the *Forty Fach* of 1884), *Fortay* 1785, *Forty Farm* 1788 etc., and in the *Diary* of William Thomas of Michaelston super Ely, *Fforddtu* 1762 and *Fortu* 1764–89 in an effort to rationalise and link the name with W *ffordd* 'way, road' and *tŷ* 'house, dwelling'.

The form is found frequently in minor names and field names in most southern English counties and is accepted as a ME form *forthay, fort(h)ey,* based on the OE adv. *forð* 'in front, before' and *ēg* 'island', the compound usually being applied to 'a ridge of higher ground projecting into a marshy area'. Although not obvious from the road, this can apply to *Forty Farm*, situated as it is on a ridge of low profile extending down into an area of low ground along the banks of the river Ely on another slope of which stands the farm of *Morlanga* (see p. 129) whose name is significant in this context.

Such an interpretation of the name is strengthened by the existence of another recorded example in the Vale of Glamorgan. The land on part of the slope of the valley of the river Kenson in the parish of Llancarfan, a tributary of the Thaw (Ddawan), near Llancadle, is also so named. It is *Forthey* 1608, *Forti, Fortye* 1622, with a reference to *The Forty Mead* by the beginning of the eighteenth century.

Gabalfa

Enquiries are still being made about this name, that of a Cardiff district on the eastern side of the river Taf opposite Llandaf. It was earlier the name of a holding and a homestead which had been divided into two portions by the middle of the sixteenth century and named *Gabalva ucha* 1527 (W *uchaf* 'higher') and *Cabalua Yessa* 1542 (W *isaf* 'lower'). Later, *Gabalfa House* comes into prominence, the surrounding area being sufficiently established by 1767 to be referred to as *Gabalfa Hamlet*.

The original location was a convenient place from which to cross the river Taf, and this is precisely what is implied in the name which appears in the *Liber Landavensis* as *coupalua super ripam taf*. It is a Welsh form compounded of *ceubal* 'boat, skiff' and the common suffix *-fa* < *-ma* 'place', *ceubalfa* being literally 'boat place', the location of a ferry, the diphthong *-eu-* > *-a-*, giving *cabalfa,* under the influence of the vowels of the second and third syllables, the whole being treated as a feminine noun, therefore lenited initially, when made specific following the W def. art., *Y Gabalfa.* The lenited form thereafter acquired permanence without the def.art. in frequent common usage.

However, this form from the *Liber Landavensis* occurs in a grant of landed property called *Villa Greguri* to an early bishop of Llandaf and which is dated by Professor Wendy Davies to *c.* 680 AD on the evidence of the names of persons named in the document. Moreover, *coupalua* itself is given either an alias or is glossed *id est penn yporth*. The whole phrase runs ... *uillam Greguri que dicitur coupalua super ripam taf . idest penn yporth* ... . In fact, the heading to the document in its printed version is *Coupalua penniporth*, the qualifying element (in modern orthography) being the compound form *pen y borth*. Its significance, in this context, arises from the fact that the meaning of the W n.f. *porth* here (the similarly formed *porth* 'gateway, door' and *porth* 'port, harbour' being masculine nouns) is 'ferry, landing-place', as in *Porthaethwy* (*Porthddaethwy*), Anglesey and *Porth Wyddno* (*Y Borth*), Cards. *Penn y porth* (with W *pen* 'the end of' in all probability), seems therefore to be synonymous with *Gabalfa*.

Ceubal is accepted as a borrowing from Vulgar Latin *caupalus, caupolus* 'small boat', OBret *caubal*, with which can be compared the English *coble* 'small flat-bottomed rowing boat' which is not considered to be a direct borrowing from the Latin and may have been adopted from the Celtic form. It can be compared with the W n.m. *cafn* 'vat, trough, conduit' also used in the sense of 'dug-out, canoe, cock-boat', and in the place-name *Tal-y-cafn* on the western bank of the river Conwy, with W *tal* 'head, end of', which denotes the location of a ferry across the river. However, *ceubal* is also used in Welsh in the sense 'belly, stomach' and Sir Ifor Williams once suggested that it acquired this meaning from what is inherent in the idea of the 'hollow' containing space of a boat.

Gabalfa occurs in Wales in the names of places other than the more well-known Cardiff example. It was the name of a grange of the Cistercian abbey of Cwm-hir in the parish of Cleirwy, Radnor, retained in the farm names *Lower* and *Upper Cabalva*, noted in Kilvert's *Diary* in 1872, earlier *Gabalva* in the *Valor Ecclesiasticus* of 1535 etc. In Sketty, Swansea, *Gabalfa Road* recalls *Gabalfa House* (*Gabalfa* 1775, *Gabalva* 1790 and, remarkably, *Ceubalfa* on the Ordnance Survey six-inch map of 1884), demolished in 1965. *Cabalfa* 1659 in Llangynidr, Brecs., appears as an element in *Tyr Maes Cae balva* 1720, *Tyr Meas Caebalva* 1732 (sic), and in north Wales, in Beddgelert, Caerns., a mansion called *Cabalfa* had its name changed to a more modest *Pen-y-bryn* in the nineteenth century.

But what of *ceubal* when it appears as a simplex form or without the suffix *-fa*? It occurs in the form *Coybal* in the parish of Llanllwchaearn, Cards. and in *Pantyceubal,* Llanfihangel Rhos-y-corn, Carms., and Cas-fuwch (Castlebythe) Pembs., *Dolyceubal* and *Coedyceubal*, Penegoes, near Machynlleth, and the old *Nant y Ceubal* (*Nant y Kybale* 1650, *Nant y gabale* 1703) in the parish of Llanedern, Cardiff. It is more than likely, as is suggested in the name *Pantyceubal*, with W *pant* 'hollow, depression, valley', that the places so named are situated in similar locations.

Galon Uchaf

A number of attempts have been made to explain this name, the most favoured being that which seeks to connect it with the old W plural form *galon*, singular *gâl* 'enemy, adversary' (sometimes 'enmity') from which is derived the current W *gelyn* pl. *gelynion* (*gâl* + the singular ending *-yn* causing affection of the penultimate vowel $a > e$). Needless to say, such a view conjures up visions of disorder and tumult in the hills around Merthyr Tudful, often quoted to justify the interpretation despite the incongruity of associating *galon* in this sense with the superlative adj. *uchaf* 'upper, higher' (lit. 'highest').

Galon Uchaf is now the name of a comparatively new urban area on the northern limits of Merthyr and to the east of a similar area, that of *Gurnos* (see p. 88), but in order to appreciate the significance of these names the nature of the community surrounding the modern industrial settlements which form the heart of the Merthyr complex should be borne in mind. It will be recalled that the main characteristic of Merthyr at the peak of its industrial development was its isolated location, like a 'dark island', as R.T. Jenkins put it, in a rural hinterland populated by tenant farmers in their scattered holdings in the hills.

David Watkin Jones in 1874 lists about a hundred of these small farms including y Goetre, y Gurnos, Bôn-y-maen, y Garn, Gwernllwyn and the like to the north of the town, one of them being named *Galon Uchaf* to which he adds *tir coed*, i.e. 'woodland'. Later suburban development occurred on the lands of some of these farms and, appropriately enough, some farm names were kept as the names of such areas. This is the case with *Galon Uchaf*.

The earliest available recorded forms of the name incline towards the acceptance of W *collen* 'hazel' as its first element to signify the location of the original homestead. The wider sense of 'young tree, sapling' is also possible but not corroborated. In addition it would appear that the second element was originally the initially lenited form of the W adj. *bechan*, a diminutive form of *bach* 'small', as a distinguishing element. The form *Callon vechan* occurs in 1740 as the name of a holding on 'the land of' (W *tir*) in *Tyr Calon Vechan* recorded in 1727, subsequently *the Collen* 1775 and thereafter *Calon ycha* 1783, *Gallon Ucha* 1784 and *Calon Echa* on Emanuel Bowen's map, 1729, with the W superlative adj. *uchaf* as the second element. If there was a 'lower' homestead, no record of it seems to be available. The form with initial *G*- shows lenition of *C*- after a lost preceding W def.art. although the evidence for it otherwise is not strong.

This evidence is reasonably substantial but it is difficult to account for the form *calon, callon* for W *collen*. If the *ll* of the 1740 form *Callon vechan* was pronounced as the W fricative consonant, the spoken variation of *-an* for *-en* in the last syllable is entirely possible and given the difficulty of non-Welsh speakers in pronouncing the Welsh *ll*, often reducing it to *l* (cf. *lan* for *llan* etc.), *colan* for *collan* < *collen* might well have been the result. A spoken vowel metathesis would therefore have to be assumed for the production of the form *calon* in *Galon Uchaf*.

Gelliargwellt

The two farms which bear this name, the 'higher', W *uchaf*, and the 'lower', W *isaf*, stand on the slope of the Waun Rydd ridge overlooking the estate of Llancaeach Fawr, south-west of Gelli-gaer. *Gelliargwellt Uchaf* is the original homestead of substantial extent and proportions and dates from the sixteenth century, if not before. It came into the possession of a branch of the Stradling family and the resident Edward Stradling, who died in 1681, had a wheat crop growing in his fields, corn in his barn, 13 milking cows, 18 sheep and a number of horses. A later resident added to the farm buildings, including a porch with a stone above the door bearing the date 1778 and the initials W.E.

A Welsh speaker today who was not familiar with the local pronunciation of the name would probably, and naturally, place the accent on the penultimate syllable, *Gelliárgwellt*, thus giving the impression that this syllable was the W preposition *ar* 'on, upon' but the available recorded forms suggest a different probability. These are: *Kellie r gwellt* 1630, *Kelly r Gwellt ycha* 1682, *Gellir-gwellt* 1697, 1885, *Gelly R Gwellt Ucha* 1842.

A clue is provided by the name of a homestead near Brogynin, Cards., to which Professor Geraint Gruffydd has drawn attention, this being *Tythin-kellie-r-meirch* 1583 which in modern orthography would be *Tyddyn Celliau'r Meirch*, with W *tyddyn* 'small holding, dwelling' and *meirch*, the pl. form of W *march* 'horse, steed' as first and last elements but having W *celliau*, the pl. form of W *celli* 'grove' as the second element (of three syllables, the accented second syllable being now marked with a diaeresis). Further, in common parlance the plural termination -*au* becomes -*a* with, in this case, the short form of the W def.art. '*r*, thus *cellïa'r*.

The problem which confronted the non-Welsh scribe was that of seeking to represent in his own orthographic medium the *sound* of the Welsh name, particularly the trisyllabic plural form *gellïau* and its affixed form of the W def.art. The anglicised form *kelly, kellie,* for W *celli* is common in documents, and the third element *gwellt* seems to have presented no difficulty, but the disyllabic ending -*ïau* in its colloquial form -*ïa* plus the affixed def.art. '*r* called for some scribal invention which was solved in the forms quoted above for 1630, 1682 and 1842 by writing the sg. form *celli* (*kelly, kellie*) + the English letter *r* to represent *cellïa'r*, as in *Kellie r gwellt* 1630.

On the analogy of *Tyddyn Cellïau'r Meirch*, Cards., what we have here, therefore, is *Cellïau'rgwellt* and later *Gellïargwellt* with lenition of the initial consonant probably through frequent use with prepositions rather than a lost W def.art. in this case. The final element *gwellt*, as seen in the compound *glaswellt* 'green grass', may well have the wider sense of 'herbage, pasture' having regard to Edward Stradling's livestock in 1681, used here as an element to signify location. It could also refer to 'straw' or 'hollow-stalked plants' like reeds etc. of the kind used for thatching and similar purposes.

Gelli-gaer

Most people will automatically regard this name as that of the large village and parish in the northern uplands of Glamorgan, the centre of the old commote of Senghennydd Uwch Caeach. It denotes, it is virtually certain, the location of a grove, W *celli*, near the site of a Roman auxiliary fort, W *caer*, on the road from Cardiff to Brecon.

However, although often used in place-names to indicate substantial Roman fortified sites as in *Caerdydd, Caerlleon, Caersws* etc., W *caer* is not exclusively used in that capacity as is sometimes wrongly assumed. Perhaps the mistaken view once held that *caer* is derived from the Latin *castra* has contributed to this notion, together with the similar use of OE *ceaster* (which *does* happen to be a loanword from *castra*), in English place-names ending in -*chester*, *Chester* itself and others like *Caistor, Caister* etc. Indeed, W *caer* is also extensively used in names which have no Roman connection, for it is a native word < British *kagro-* which has the basic sense of 'enclosure', and is closely related to the noun *cae*, now 'field' but earlier 'a hedge, a barrier' (cf. the W *cae drain* 'thorn hedge'), i.e. that which encloses, subsequently the land so enclosed to form a field. It is cognate with OE *hecge* 'hedge'. Its use in Welsh, therefore, to signify hill-forts and the like which are themselves basically 'enclosures' is rational. It is a feminine noun, its initial consonant being lenited in asyntactical compounds when preceded by an adjective as in *(Y) Gron-gaer* 'the round *caer*' (W *cron*, f. form of *crwn* 'round, circular') and *(Y) Gelli-gaer*, where *celli* has a qualifying function.

Other places named *Gelli-gaer* in Glamorgan, in none of which does *caer* refer to a Roman fortification, support this fact. They are farms whose names were acquired when they were larger units before subdivision at a later stage.

Gelli-gaer Fawr and *Gelli-gaer Fach* stand on the slopes of *Mynydd-y-gaer* to the east of Briton Ferry (Llansawel), a subdivision of an earlier single holding, *Kilticar* 1336, *Kelli y Gaer* 1538, *Kelly yr Gare* 1611 etc. The *caer* here is the Iron Age hill-fort on the summit of the hill which has its own name *Buarth-y-gaer* and which has inside its enclosure an earlier Bronze Age cairn. The use of W *buarth* in the sense of 'an enclosed space' (now used mainly in relation to farm locations in the sense of 'a fold, penfold, enclosure' for milking cattle or penning other animals) as the name of an ancient enclosure which is virtually circular in shape preserves the basic sense of *caer*.

Further, in the parish of Cilybebyll, there is a name which is no longer current but is recorded from the sixteenth century onwards in a sequence of the estate papers of Plas Cilybebyll: *Kelli Kayre* 1520, *Kelli y Kairei* 1535 (with the plural form *caerau*), *Kelli'r Kayre* 1536 etc. In 1650 *Kilybebyll ysha alias Kelli Kayre* is recorded (*ysha* = W *isaf* 'lower') but the location of the property is not certainly known. Existing documentation will not allow its identification with the existing ruinous *Cilybebyll Fechan* at National Grid reference SN 7444 0523 with any confidence despite the temptation to assume a broad correlation between the qualifying adjectives *bechan* 'small, lesser' and *isaf*. Furthermore, an added complication is the possibility of this property having acquired another name, for in 1813 reference is made to *Ton-y-fattu anciently called Kelly-Cayre*, and there is further evidence for this holding, in turn, to have been subject to subdivision: *Ton y devatty ycha* and *-issa* 1520, *Tonne y divattie isha* 1556-7, *Tone y dyvattye ychae* 1557, *Ton y devattye* 17c., *Tonydufattu* 1810, *Ton-dy-fattu-ucha* and *-issa* 1813. This is the W n.m. *ton* which has the primary sense of 'skin, surface' but occurs extensively in Glamorgan place-names for poor, unbroken, unploughed land or layland, pl. *tonnydd, tonnau* (> *tonna* in the vernacular, as in *Tonna*, near Neath) + *y* + *dafaty* 'sheepfold, sheepcote' < *dafad* 'sheep' and *tŷ* 'house' (as in *Dyfaty*, Swansea). In this connection, I am grateful to Jeff Childs for drawing my attention to the fact that one plot of land on the farm now named *(Y) Wigfa*, to the east of the church of Cilybebyll is named *Cae de fatty* ('the Dyfaty field') in the tithe apportionment for the parish in the 1840s. No authentic remains which could be associated with W *caer* are noted in this vicinity by the Royal Commission on Ancient and Historical Monuments.

In all three cases, but more particularly in the case of *Gelli-gaer* in Senghennydd Uwch Caeach which is far more comprehensively documented than the others, the collected forms of the names from a comparatively early date include a substantial number which suggest that the scribes, more often than not, recorded the name in the form in which it was commonly pronounced in the vernacular, a form which persists to this day. This appears, for example, as *Gelli Gare* 1739, to rhyme with E *bear, dare, air* etc. and appears to be what was intended to be conveyed by forms like *Kilthegayr* 1281, 1306, *Kilthigair* 1314, *Kylthergayre* 1376, *Kilthigayr* 1441, *Kelly Gaire* 1491 etc. This follows the normal Welsh Gwentian dialect pattern of first reducing the diphthong *-ae-* in the second element to a long *-ā-* (*Kilticar, Killecar* 1208, *Killigar* 1567 to *Gelligar* 1738, 1740) which was, in turn, narrowed to *-ǣ-*, so that W *caer* > *cār* > *cǣr*, reproduced in the initially lenited form *gare* (cf. p. 87). The simplex form *Y Gaer* itself is nearly always pronounced as *Y Gare* in more than one location in Glamorgan.

Gendros

The name of a well-known Swansea district and like others in the area such as Hafod, Penfilia, Cwm-du, Pen-lan, Tirdeunaw etc. highly suggestive of non-urban origins. Such areas were once open country, hills and dales, woods and moorland.

Gendros is composed of two elements, W *cefn* 'back, ridge' + the

W n.f. *rhos* 'moor, heath'. In common with several other names which have *cefn* as a first element there is a tendency in common parlance to lose the *-f-* in the combination *-fn-* in an accented syllable. A similar example is Cefn-faes near Llantwit juxta Neath (Llanilltud Nedd) which has W *maes* 'open field' as the second element and became *Cenfaes* in local speech, recorded as *Kenfas* in 1859. The fairly common Cefn-coed with W *coed* 'wood, forest' as second element, became *Cencoed*, with its variant form Cyncoed in Cardiff (but not in Swansea) and Kingcoed elsewhere.

Similarly, by 1735 the Swansea Cefn-rhos had become *Cenros* (*Genrose* in the hand of an English clerk in that year), and by 1844 *Genrhos* and *Genrhos Farm* are recorded where the initially lenited form shows the influence of the Welsh definite article used to make the name specific, *Y Genros*, originally in reference to a holding or farmstead so named because of its location, in all probability. In the name of *Cenrhos farm* on a southern shoulder of Mynydd Pen-bre and north of Porth Tywyn (Burry Port), Carms., on the other hand, no lenition is evidenced.

The name was to become subject to a further colloquial change, namely the growth of *-d-* in the combination *-nr-*. This follows the pattern of development of names like Penr(h)yn being pronounced Pendryn, the Welsh personal name Cyn(w)rig as Cyndrig (anglicised Kendrick), and words like *cynr(h)on* 'worms' becoming *cyndron* etc., so that Y Genros is recorded as *(Y) Gendros* by 1846, *Gendros Farm* 1852, and retained in that form as the name of the surrounding district.

It is relevant to note that a form in *-ndr-* can cause the misinterpretation of a name if viewed superficially. As Dr B.G. Charles has shown, the Welsh form of Henry's Moat in Pembrokeshire is Castell Hendre which lures the unwary to see in it the well-known Welsh term *hendre(f)* as the second element, but as the basic English form indicates, it is the personal name Henri, Henry with the intrusive *-d-*, giving *Hendri*, and with its loosely pronounced unaccented final syllable *hendre*, with W *castell* 'castle' for the E *moat*, which is ME *mote* 'embankment, mound' here rather than, simply, a surrounding ditch.

Y Gnol : The Knoll

Probably better known to aficionados of rugby football as the name of the ground on which the Neath club play their home fixtures, The Knoll is the common English word for a mound or hillock. From the fourteenth century at least, the town of Neath being a plantation borough, most of the early names of its streets and close surrounding natural features were English and a hillock on its eastern extremities is recorded as *La Knolle, lez Knoll* in 1570 and *The Knoll* on Emanuel Bowen's map of 1729. This was the feature on which the family of David Evans of the Great House in the town raised a substantial residence and it was into this family that one of the daughters of the industrialist Sir Humphrey Mackworth married. The house acquired the name and was demolished in 1956, the Neath rugby club ground being part of the estate.

However, by the sixteenth century there is evidence that a Welsh form of the name, *Y Gnol*, had come into circulation. The antiquary Rice Merrick refers to a stream *nant y gnoll*, which ran by the hill, in 1578, and references to *Cae'r Gnoll*, *Heol y Gnol* and *Penhewl y Gnol* occur in 1666, 1684–5. In fact, the vast majority of documentary forms of the name collected dating from the seventeenth and eighteenth centuries are in this form rather than *knoll*. The interest, of course, lies in the fact that the initial consonant in the English *knoll* is silent, whereas in the Welsh form not only is there a sounded consonant but a consonant which appears to have undergone the process of lenition of an original *c*-, (*k*-), after the definite article in the Welsh manner, *Y Gnol*.

Although the initial consonant in the modern English *knoll* is silent, it derives from the OE *cnoll* 'summit of a hill' in literary OE but 'liable to have been applied to any small hill in names of comparatively recent origin', according to Margaret Gelling. It was a form, however, in which well into the Middle English period the initial *c*- was sounded. Several Welsh words beginning with *cn*- were borrowed from English during this period and they preserve the original pronunciation: W *cnoc* < OE *cnoc(ian)*, ModE *knock*; W *cnap* < OE *cnæpp*, ModE *knap*; W *cnaf* < OE *cnafa*, ModE *knave* etc. and the name of a prominent mountain in Snowdonia, Y Cnicht, is a borrowing from OE *cniht*, ModE *knight*. This does not seem to have happened in the case of OE *cnoll* except, perhaps, in the name discussed here, and it is suggested that in the Neath area the local inhabitants were accustomed to hearing the initial *c*- (*k*-) being sounded in the medieval period and accepted it in common Welsh parlance as a feminine noun with subsequent lenition of its initial consonant after the Welsh definite article (cf. the treatment of the second element of the name Tre-os, p. 201).

It is difficult to accept Rice Merrick's form in any other way. Again, it must be borne in mind that this was essentially an oral process, not an orthographical exercise, a good comparable example being the name *Y Gopa* in the parish of Llandeilo Tal-y-bont borrowed from ME *coppe* 'a summit, top, mound' and in Merrick's day the Welsh form *gnol* was undoubtedly pronounced with the full value of the initial consonant.

A final twist to the story is that by today even the initial *g*- in the Welsh form Y Gnol is not enunciated; it is silent, as is the English *k*- in The Knoll. It may well be that the modern influence of that group of English words in *gn*-, like *gnarl*, *gnat*, *gnome*, *gnaw* etc., which are initially pronounced in similar fashion has come into play.

Gowlog

It is reasonably clear from the recorded forms of the name of this farm near Llanfeuthin in the parish of Llancarfan that the diphthong –*ow*- in the modern form is an anglicised representation of the Welsh diphthong –*aw*– in an original form *Gawlog*, the initial consonant being the lenited form of an original in *c*– after a lost def.art. used to make the name specific, Y Gawlog, and as in the majority of Welsh farm names, thereby indicating its feminine gender.

These are the earliest forms which have been collected: *The Iawloge Lands* 1657, *Gawlog* 1672, *Gawlogg* 1673, *Gaulock* 1728, *The Gawlock* 1753, subsequently *Gowlog* 1764, 1767 etc. The elements are W *cawl* + the W collective pl. ending –*og*, MW –*awg*, –*awc*, which has the function of denoting a place where an abundance of the plant referred to in the first element is to be found.

W *cawl* is, therefore, not to be interpreted here in its modern sense of 'a dish of vegetables, alone or with meat, boiled to produce broth, soup or pottage' but in its original sense of a species of brassica, probably 'wild cabbage', derived from the Latin *caulis*, as in the first element of E *colewort*, *cauliflower* and *kale* (see pp. 216–17). *Y Gawlog* would have been named after a location abounding in such growth, and in structure the form is similar to other names like Y Fanhadlog (W *banadl* 'broom'), Y Gelynnog (W *celyn* 'holly') compared with Banadlog, Banalog, Clynnog, which have a wider adjectival connotation.

An alternative way of indicating the meaning which forms in *–og*, *–awg* convey in W names is the use of the W def.art. + the lenited form of the name of a particular plant without the addition of a collective plural ending, as in Y Rug (W *grug* ' heather'), Y Wern (W *gwern* 'alder'), Y Gors (W *cors* 'reeds'), the last two, in particular, having the further developed sense of 'marsh, swamp', indicating the wet and damp environment in which the plants flourished. On this pattern, a form such as *Y Gawl* might have been possible in this case but is not recorded with certainty, the form *Tyle'r Gawl* (W *tyle* 'slope, ascent, hill') quoted by Iolo Morganwg in connection with this broad area having been considered by the late Professor G.J. Williams not to have any connection with this particular location.

I am informed by Dr John Etherington, editor of the Llancarfan Society's Newsletter, that the cliffs on the coast to the south of the parish are yellow with the flowers of wild cabbage in spring. He also adds documentary evidence from the 1890s to the 1930s to the effect that in local parlance the name Gowlog took upon itself the further interesting popular guise of *Gold Oak*. This does not seem to have survived.

Groes-faen

W *croes* 'cross' can occur in place-names either in a substantival or an adjectival capacity, the latter in *croesheol* (with W *heol* 'way, road, street') or *(y) groeslon* (with W *lôn*, borrowed from OE *lone*, a variant form of *lane*) both meaning 'crossroads'. The location so named is sometimes referred to in the substantival simplex form Y Groes, especially in the compound Pen-y-groes, and in the parish of Llanrhidian in Gower the plural form *crwys* is also used (but see also p. 97). As a noun it is usually feminine, its initial consonant being lenited after the def.art. although an example is quoted in GPC of an unlenited masculine form, *Y Crôs*, for the main crossroads at the centre of Morriston, Swansea. Here, the typical southern reduction of the diphthong *–oe–* to the long vowel *–ō–* has occurred in the vernacular.

As a noun, the reference is usually specifically to a cross, of stone or wood, sometimes a wayside cross or calvary, sometimes a boundary cross. The boundary between the parishes of Pen-tyrch and Llantrisant runs through the crossroads in the centre of the village of Groes-faen, a name also borne by a farm, and the second element, *maen* 'stone', indicates that the reference is not to the crossroads in this instance but to a cross which must have stood nearby. Normally, W *maen* is also a substantive used in place-names like Maenclochog, Maentwrog etc. but is likewise used extensively as an adjective as in the present case, 'stone cross': *Croyse Vaen* 1570, *y groes faen*, *Croes vaen* 1630, *Croes Vane* 1636 etc., the latter form beginning to reflect the pronunciation of the name in common speech, having not only the reduction of the diphthong –*oe*– to –*ō*– in the first element, as noted above, but also the similar reduction of –*ae*– first to a long –*ā*– and then, a feature of the Gwentian Welsh dialect, the 'narrowing' of the long –*ā*– to –*ǣ*– to give a form which is represented by English scribes in their orthography as –*vane*, thus *Grosvane* 1650, *Crosvane* 1783, *Crossfane* 1768 etc. with which may be compared the form *Lisvane* for Llys-faen, Cardiff, or the spoken form *Llanmace* for Llan-faes in the Vale of Glamorgan.

Confirmation of the function of the cross in this location comes from a single English form recorded in 1492, *Harston*. Although no other examples of this form are available it can be clearly recognised as a form which has the variants *Hoarston*, *Horston*, *Whor(e)ston* and the like found extensively in England with examples also in Pembrokeshire and at least three in Gower. This is OE *hār* 'grey, lichen-covered' + OE *stān* 'stone', *hoar-stone* in ModE, which acquired the sense of 'boundary-stone'. Thomas Richards in his Welsh dictionary (1753) defines *croesfaen* as 'a mearstone marked with a cross'. Whether this is exactly true of the example noted here or whether an earlier boundary stone was later replaced by a cross is not known.

(Y) Groesfaen occurs as the name of a farm in the parish of Gelli-gaer, as the name of a commote in the old Powys Fadog and, according to GPC, the name of a stone on the boundary between two townships in Tywyn, Merioneth. There are probably many more.

Y Gurnos

The place of this name which makes the media headlines most often for the wrong reasons is the well-known suburban area of Merthyr Tudful, but it is not the only example of the name in Glamorgan. Gurnos occurs near Gowerton, also Bryngurnos (with W *bryn* 'hill') in the hilly region between Maesteg and Cwmafan. Further afield, just over the northern border, is Gurnos near Ystradgynlais and Gyrnos in the parish of Llanafan Fawr, both in Breconshire, to the east of Llandeilo, Carms., and on the north western edge of Brynmawr, Mon. Gernos Mountain is in the parish of Llangynllo, Cards., and a prominent height, Pen y Gurnos stands in the Doethie valley to the west of Llyn Brianne in the same county.

The Merthyr example was first the name of a farmstead, *gyrnos* 1630, *y Girnos* 1716, *y Gwrnos* 1794, *the Gurnos* 1815, *Gyrnos* 1874, and a water grist mill, *Melyn y Girnos*, is recorded in 1716. The farm was clearly named after a feature of the land on which it was raised, later to become that of the urban area which developed on this land as was the case with Galon Uchaf (p. 78).

It is a name which is difficult to explain satisfactorily and one which caused some uncertainty even to a lexicographer of the stature of Sir Ifor Williams. It is, nevertheless, in the light of his discussion of the form *gurnos* that any subsequent deliberation must be attempted. The basic form is the n.f. *curn*, *cyrn* (both these forms appear indiscriminately) which is defined in the earlier dictionaries as 'heap, pyramid', or *acervus*, *pyramis* by Dr John Davies of Mallwyd (1632), to which GPC adds 'mound, cone, rick'. The prime examples of its use are the names of the mountains Y Gurn-goch, Y Gurn-ddu, and Y Gurn-las in Caerns., the three being cone-shaped, which is taken to make it unlikely that *curn*, *cyrn* here is the plural form of W *corn* 'horn' used in the sense of 'a cairn or point on a mountain top'. The plural form Y Cyrniau is evidenced as the name of a group of hills elsewhere. Neither did Sir Ifor accept that the lenited initial consonant after the def.art. was an indication of the old Welsh dual plural form because the three examples of the names in Caerns. refer to individual mountains, not to two in each case. See also Pen-cyrn (pp. 141–2).

The exact function of the ending *–os* is not entirely clear in this context. Normally, it is a diminutive ending, as in *plantos*, *merchetos*, *gwrageddos* etc., which developed a collective plural sense, particularly when attached to the names of plants and vegetation, to indicate a place where an abundance of such growth was to be found, as in *bedwos* (with W *bedw* 'birch', the origin of the place-name Bedwas), *brwynos* (with W *brwyn* 'reeds'), *grugos* (with W *grug* 'heather', as in Rhigos which is *Grygos* 1570, *y rygos* after 1590 etc., locally (*y*) *Ricos*) in the northern uplands of Glamorgan, and near Tonna, Neath (*Tone Gregoys* 1527), also in the Dulais valley in the present form Glynrhigos but originally *Clun y Rhigos* (*Clyn y Rigos* 1666) and several other places. It is difficult to see how appropriate such an ending would be in the forms *gurnos*, *gyrnos*, since *curnen*, *cyrnen* are the normal diminutive and singular forms of *curn*, *cyrn*.

Where W *bryn* 'hill' occurs in a compound with *gurnos*, *gyrnos*, as in the Bryngyrnos noted above, or where a name like Pen y Gyrnos, Cards., occurs in hilly surroundings as do other simplex forms like Y Gyrn in Glyn Tarell, Brecs., there can be little doubt about the meaning of *curn*, *cyrn*, the ending *–os* having been added, possibly, to convey some sense of diminution. In other locations, such as the ones near Gowerton or Merthyr, it is difficult to accept a prominent hilly or mound-like situation, conical or otherwise. B.G. Charles notes Gernos in the parish of Nevern, Pembs., which has the forms *Girnos* 1714 and *Gurnoss* 1778 among others, but which he thinks may be for W *gwern* 'alders' + the diminutive *–os*, and Gernos in the parishes of St Dogmaels, Llanfyrnach and Meline in the same county. Perhaps the most acceptable solution for the Merthyr Gurnos may be to regard the name as a descriptive term in such a location for land whose uneven surface is characterised by tumps or hillocks and depressions.

Gwernycegin

The main road from Cardiff to Llantrisant runs down the slope of a ridge to Rhydlafar (p. 167) and in periods of prolonged wet weather its surface is covered with water which overflows from a wet and marshy wooded expanse on the north side of the road. This justifies the existence of W *gwern* as the main element of the name of that area, Gwernycegin.

Gwern is a collective plural form of *gwernen* 'alder tree', and Sir Ifor Williams once reminded us that in place-names the use of this form, lenited initially after the W def.art. *Y Wern*, has a similar function to that which occurs when plant or vegetation names have attached to them the collective plural ending *–os* (pp. 88–9) to denote a place where an abundance of the named variety is to be found. However, some of the forms which are in the same category as Y Wern, the most common being Y Gors, where *cors* is a plural form of *corsen* 'reed, reed-grass, stalk, stem', have themselves acquired a derivative sense as the term used for the ground which is the normal habitat of the named plant. Thus *cors* now means 'marsh, swamp, bog' and *gwern* has acquired a similar sense, being the kind of ground on which alder-trees flourish, 'marsh, swamp, damp meadow'.

What of the second element, *cegin*, with the W def.art. *y cegin*? That it is not the common W n.f. *cegin* 'kitchen' is proved by the fact that the initial consonant is not lenited after the def.art. It is, rather, a word which is not now in current use but which survives in some place-names and is a variant form of *cain, cein(g)* 'ridge, back, hog's back'. It can also be found in a number of old ecclesiastical property boundaries in the *Liber Landavensis* in the OW form *cecin, cecyn* as in *cecin penn icelli* (ModW *cegin pen y gelli*), *cecin iminid* (ModW *cegin y mynydd*), *cecyn crib iralt* (ModW *cegin crib yr allt*) and *cecin meirch* 'the horses' ridge' in Llandeilo Fawr, Carms., which corresponds to the second element of the name Llanrhaeadr yng Ngheinmeirch, Denbs. (sometimes wrongly spelled Cinmerch). Carnau Cegin and Cerrig Cegin, Carms. are noted in GPC as names which contain the same element.

What we have in Gwernycegin, therefore, is marshy ground on a ridge of land which seems, on the grounds of having the archaic form *cegin*, to have a measure of antiquity but for which no satisfactory early evidence has been found to date. *Wern y cegyn* occurs in 1833 and *Gwern y Cegyn* in 1884. The latter form, however, also appears on a Plymouth estate map of 1766, a feature of which is that the surveyor has attempted to give bilingual Welsh and English forms for minor names and field names. Gwernycegin is translated as *Kitchen Arles*, with *kitchen* erroneously for *cegin* in this context, as noted above, and *arles*, the ME form of *alders*.

Cegin also occurs as the name of a stream which flows into the Menai at Abercegin (Port Penrhyn), Bangor, Caerns., possibly so named from its source in Ffynnon Gegin Arthur in the parish of Llanddeiniolen. This was the view of R.J. Thomas who also believed that the lenited form *gegin* in the name of the well implies that the sense of 'kitchen' is appropriate in that context, basing his belief on a tradition that the well was a kind of cauldron associated with the legendary Arthur because of a mineral scum which collected on the surface of the water with an accompanying gaseous aroma which was likened to 'cooking' smells. I am informed by Tomos Roberts, however, that fairly early evidence exists for a variant form, *cegid*, as the name of this stream (W *cegid* having the normal sense of 'hemlock'). Usually, the prevailing variant form of *cegid* is *cegyr*, and it is suggested that this is the stream name to be seen in Abercegyr, Mont. Further, we are informed in GPC that a colloquial form of *cegyr*, namely *cegyrn*, is current in Cards.

Though it is interesting to see Gwernycegin presented as *Gwern y Cegyrn* on an earlier Ordnance Survey 25-inch map of the area, it may be unwise to see in this form anything more than the acquisition of an intrusive *–r–*, whatever may have been the cause of its appearance.

Y Gyfeillion

In his series of essays on the history of the parish of Llanwynno written well over a century ago, William Thomas (*Glanffrwd*) maintained that at the time he was writing (Y) Gyfeillion, between Pontypridd and Trehafod, near Porth, Rhondda, was already an old village. Early recorded forms of the name, however, do not seem to be plentiful and the few that have been collected do not show much variation in form except for the common tendency in south Wales to lose the consonantal *–i–* in the final syllable *–ion*, thus producing the colloquial form Gyfeillion.

With the help of the late R.J. Thomas's notes we can recall that behind the village the ground rises very sharply but with a break running northwards between two steep rocky slopes leading to a narrow valley. One Welsh word for such a slope is *allt* (it can also mean 'a wooded slope' in south Wales) and its modern plural form is *elltydd*, although the more obvious but less common *alltau* does occur, as in Penalltau, the name of a farm which became the name of a well-known colliery near Ystrad Mynach. These, however, are fairly late forms, an earlier mutated plural form being *aillt* which, further, had a separate old dual plural form (to signify two of something), *eilltion*.

On the other hand, another way of expressing in Welsh the

existence of two things in close proximity or which had a common existence or similarity of form or appearance was to use the prefix *cyf–* < Celtic **kom–* before their names as was the case with W *ynys* 'island, water-meadow' as the basis of several examples of Y Gyfynys in Caerns. Also to be noted is the lenition of the initial consonant after the definite article, not as a feminine noun in this case, but as an indication of a dual plural form.

If we go back to the dual form *eilltion* noted above and postulate a form with the prefix *cyf–* which intensifies its dual sense, this would produce the form *cyfeilltion*, this becoming *cyfeillion* in common usage with loss of *–t–*. This is entirely possible when it is recalled that the plural form of MW *cyfeillt*, ModW *cyfaill* 'friend' (but differently derived) was also *cyfeilltion*, then *cyfeillion*. With the definite article and initial lenition to indicate the dual sense we have Y Gyfeillion 'the two facing rock slopes' as a place-name.

Locally, it is known that the south Wales propensity for reducing a diphthong to a lengthened form of its dominant vowel would see the *–ei–* of Gyfeillion becoming *–ī–* to produce the form pronounced as *Gyfillon*, and it is suggested that this is also the basis of the form of the name which now appears as Govilon between Abergafenni and Gilwern, Mon., (*Govylon* 1689, *Gofeilion farm* 1798), the W fricative *–ll–* again being reduced to *–l–* by non-Welsh speakers as in the case of Galon Uchaf (see p. 79). Govilon also stands between the steeply sloping sides of the entrance to Cwm Llanwenarth.

Hafodhalog

The name of one of the granges of Margam Abbey which is still the name of a farm to the north of Mynydd Cynffig but which rarely appears on maps in this, the correct form.

It is a name which has been consistently misrepresented by scholars and historians over the years as *Hafodheulog*, as if the two elements of which it is composed were W *hafod* 'summer residence' and the W adjective *heulog* 'sunny, sun-bathed' (W *haul* 'sun' + the adjectival suffix *–og*). This is also the form favoured by the mapmakers, so that perhaps it cannot be claimed that it is entirely the result of popular etymology. Possibly, a better description would be to call it a fanciful creation.

At least we are helped here by a very extensive selection of earlier dated forms of the name, a substantial number dating from the thirteenth century, among them *Hevedhaloc, Havedhaloc, Hevedhalok, Hauodhalawc, Havethaloc, Havothaloch*, and although there are numerous variations, particularly in the first element of the name, which are obvious examples of non-Welsh scribal efforts to cope with *hafod*, there is a remarkable consistency in the presentation of the vowel *–a–* in the penultimate syllable. There is little doubt that the second element is the W adjective *halog*, not *heulog*. Over fifty early forms of the name are attested, and not one has a trace of the diphthong *–eu–* in the penult.

This presents no difficulty where the meaning of the name is in question for in fact it confirms some of the documentary evidence. W *halog* is an adjective based on the substantive *hal* 'moor, marsh, moorland' + the adjectival suffix *–og*. Some of the references in the earliest thirteenth-century Latin documents (1200–14, 1208, *c.* 1213–16) are to the *mora* of Hafodhalog between the rivers Cynffig and Baeddan, a notable feature of the vicinity. Far from drawing attention to the sunny aspect of the *hafod*, therefore, its name emphasises its marshy surroundings.

The W adjective *halog* occurs also in the derived sense of 'dirty, foul, muddy, miry' and is cognate with Irish *salach*, of similar meaning. Used substantivally as a diminutive form *halogyn*, it occurs in five or six instances as a stream name in Wales, usually in a reduced colloquial form as in Logyn, Login, near Waunarlwydd, Swansea.

Numerous also are names in Wales compounded of W *rhyd* 'ford' + *halog* to denote a well-used miry, muddy ford, Rhydhalog, and with medial provection of –*d* adjacent to *h*– in the form Rhytalog (see Nant Talwg, p. 56). Plwcahalog occurs in Cardiff (*Pluccahalog* 1774) with W *plwca* 'plot of land' as first element, and Wernhalog (*Wernhaloc* 1417) with W *gwern* 'marsh, bog, swamp' near Llanrhidian, in Gower.

One of Edward Lhuyd's correspondents, writing *c*. 1700, notes the name of one of the substantial houses of Merioneth as Yr Havod Talog in the parish of Llandecwyn. Here, unlike the Glamorgan example noted above, the medial *d*+*h* consonantal combination has resulted in the provection noted in the form Rhytalog, but the marshy surroundings of the Merioneth property is also unmistakenly corroborated by a further reference in Lhuyd's *Parochialia* to Gwern yr Havod Talog in the same parish.

Hafodwgan

On a high shoulder of land on the borders of the parishes of Margam and Llangynwyd which overlooks Cwm Cynffig on the west and the source of Nant Craig yr Aber on the east there stood another of those dwellings which were characteristic of the practice of transhumance as part of the annual cycle of stock-rearing activity in Wales. It was another *hafod* and an outlying portion of the Margam Abbey grange of Hafodhalog. However, a search for it by the name which appears as the heading to this note would prove fruitless, its form on present Ordnance Survey maps being Hafod Decca, for Hafod Deca, this in turn being the colloquial form of Yr Hafod Decaf, having the superlative form of the W adj. *teg* 'fair, pleasant', *tecaf*, as second element in an attempt to rationalise a form that seems to have given scribes from the thirteenth century onwards some difficulty. Furthermore, what remain today of the farmstead are the ruins of the walls clustered around a small yard being rapidly overgrown by dense forest.

Scribes have presented the form as *Havod y Dyga* 1527, 1633, 1700, *Havod y Diga* 1681, 1713, *Hafod Digoed* 1814 and even *Hafod Beca* 1833.

In the Papal Taxation of 1291 there occurs a reference to 4 acres of land valued at 6d. at *Handugan*, a form which would be difficult to fathom were it not for the existence of a variant reading, *Kavodduga* (where *k–* is an error for *h–*) and which is corroborated by the form *Havothduga* in an extent of Margam land in 1336. Even so, the reading *Handugan* is not to be rejected entirely if the first *–n–* is read as *–u–*, *Haudugan* (the similarity between *n* and *u*, letters of two minims, being well known in early documents). Likewise, *Haudugan* is a likely erroneous form of *Hau(o)dugan* by loss of the medial *–o–*. Since *u* and *v* are also similar in manuscripts we have *Havodugan* as the probable base for the 1291 *Handugan*. This form also suggests the presence of a terminal *–n* in the original form of the name.

Evidence exists in the Melville Richards Archive at Bangor for the place-name *Hafod Wgan* 1543 in the parish of Llanedi, Carms. Further, in a charter of the monastery of Strata Marcella (Ystrad Marchell) there is a reference to a *Hauot gwgaun* 1206 in Powys. These names clearly confirm the presence of the old W pers.n. Gwgawn as the second element which later becomes Gwgon and Gwgan, as evidenced in literary and toponymical contexts and with which the name noted here appears to be identical.

The lenition of the initial consonant of Gwgan in Hafodwgan follows the Welsh pattern when personal names are attached to feminine nouns in a genitival relationship in compounds to show occupation or ownership, as in Rhyd Wilym, Hafod Ruffudd, Hendre Forgan etc.

This interpretation is preferred despite the fact that the form Gwgan is also the second element of the W pers.n. Cadwgan and may occur, by loss of the unaccented first syllable, in the name of Tredogan, a hamlet near Fonmon in the parish of Pen-marc. This can be compared with Trecadwgan, Pembs., which seems not to have been shortened in popular usage over the years and in any case is what B.G. Charles called 'a translated name', the earliest recorded forms being *Cadygan(i)ston* 1326, *Cadyganeston* 1363, similar forms being unavailable for Tredogan.

Henstaff

At another location on one side of the main road from Cardiff to Llantrisant, between Groes-faen and Capel Llanilltern, and at the end of a short driveway, there stands a substantial house now known as Henstaffe Court. Built in the nineteenth century, it is close to the site of an original farmstead, later a gentry house, as J. Barry Davies has demonstrated, standing on land assarted from parkland described in a survey of 1570 as 'forest replenished all in okes'. This was the old Parc Coed Marchan (now the name of a farm) straddling the parishes of St Brides super Ely, Peterston super Ely and Capel Llanilltern and which is to be connected with the *mainaur cruc marchan* of the *Liber Landavensis* in a grant which has been dated to *c.* 1040 (a farm called Pen-crug stood nearby).

Henstaff was probably referred to as *Tir-y-bryn* in 1570, but the first record of the present name seen hitherto is *Henstab* which, although occurring in a late seventeenth-century MS copy, is in the work of the Glamorgan poet Dafydd Benwyn who flourished in the second half of the sixteenth century, in this case an elegy on Jennet (*Siwned*) Mathew, wife of Richard Harry of Henstaff. Subsequent forms are *Henstapp* 1600, *Henstaffe* 1607, *Henstappe*, *Henstod* 1608, *Henstabb* 1706–7, *Henstaff* 1776 etc. and as such regularly in eighteenth-century rent-rolls coupled with Tir y brin, Tir y Bryn 'the hill land'. However, the form *Henstabl* occurs in an undated document, possibly of the eighteenth century, and this form is confirmed by *hen stable c.* 1625 in a Bute rental which is remarkable for the fact that it is written entirely in Welsh.

There is no reason to doubt that the elements of which the name is composed are W *hen* 'old' + W *stabl* 'stable' (*stabal* in north Wales) probably borrowed from ME *stable*, though it will be recalled that the English form is borrowed in turn from OFr *estable*. In common speech *Henstabl* > *Henstab* by loss of final *–l* in the group *–bl* as occurs in W *posibl* > *posib* (or in a *–gl* grouping as in W *perygl* 'danger' > *peryg*, *perig*) with further provection of the final consonant in the form *Henstapp*, *Henstappe*.

The form *Henstaff* is first evidenced in 1607 and appears to be a variant of *Henstab*, on the evidence of the pronunciation of W *cwnstabl* 'constable' (borrowed from ME *cun(e)stable*) as *cwnstaff* in Dyfed (see GPC) since there is evidence of the intermediate form *kwnstab* in Peniarth MS 177 (written 1544–65).

Heol-y-crwys : Crwys Road

There is evidence for the use of the form *crwys* as the plural of W *croes* 'cross' from the fourteenth century but where its etymology is in question it is borrowed from the singular Latin *crux*. What seems to have happened is that by analogy with W plural forms like *ŵyn* 'lambs', singular *oen*, or *crwyn* 'skins, hides', singular *croen*, *crwys* came to be regarded as a plural form for which *croes* was adopted as the singular, although the reservation is made in GPC that it could be a borrowing from a hypothetical Latin form **crox*. The plural form *croesau* is a later creation by the addition of the common W plural ending *–au* to *croes*.

It is not always clear whether *crwys* is to be understood in the singular or plural sense when it occurs as an element in place-names unless there is additional corroborative evidence. For Y Crwys near Penclawdd in Gower, the E form *the three crosses* is evidenced 1707, 1729, 1731 etc. and may refer to old wayside crosses or boundary crosses. This may well be the case with Pant-y-crwys, Craig-cefn-parc near Clydach, and what of Llanddewi'r Crwys (Llanycrwys) near Pumsaint, Carms?

Further, *crwys* is usually regarded as a singular n.f. in the dictionaries. A text of 1595 contains a reference to *y grwys* on which Christ was crucified (with which may be compared *y groes*), but there is room to doubt that this is invariable. In Heol-y-crwys (Crwys Road), Cardiff, documentary evidence provides firm grounds for regarding *crwys* to be not only a singular form but also that it is a masculine noun in that context—the initial consonant is not lenited after the definite article, *y crwys*. The street name preserves the name of a holding, later a farmstead, which was on the land of a grange of Margam Abbey in the Middle Ages and which stood near the northern end of that street, namely Y Crwys Bychan, the qualifying element being the masculine form of the adjective *bychan* 'small, lesser'.

Some of the earlier forms are *Crosse byghan* 1540–53, *cross bayghan* 1675, 1682, *crugbogan* 1722, *Crossbughan* 1731, *Crusbuchan* 1732, 1736 etc. to *Crwys bychan* 1824, *Crewis Buchan* 1828. The old farmstead stood on the northern boundary of the parish of St John, Cardiff, and its name probably indicates the location of the boundary cross after which it was named. The farmhouse was demolished in 1899 and the present Gladstone Schools were erected on the site. Some way to the east, in the parish of Roath, there once stood the farmstead of Y Crwys Mawr (with W *mawr* 'large, great').

Heol-y-march

This was the name of a smallholding where today a couple of small houses stand by the side of the road which surrounds the farm of Maes Siward (erroneously shown as Maes-y-ward on the Ordnance Survey maps) near Welsh St Donats in the Vale of Glamorgan.

As it stands it appears to be a simple name composed of the Welsh elements *heol* 'street, way, road' + the W definite article *y* + *march* 'horse, steed', thus 'the horse's way', which seems quite appropriate for an age uncluttered by modern motor vehicles. In addition the word *heol* is not far removed from what was intended in the original form.

During the course of time, the most fundamental change has been in the form of the name rather than its meaning: *Ollmarch* 1603, *Olemarch* 1612, *Olmarch* mid 17c., *Olmargh* 1628 etc., including *Old March* 1659 by a non-Welsh scribe.

The first element is W *ôl* 'impression, imprint', hence 'path, track' not commonly found in the modern period in that specific sense, which has the plural form *olau* noted in the discussion on Uchelolau and Rheola (p. 210). Heol-y-march developed from a colloquial form of *ôl (y) march* 'horse's path, or track' and is evidenced in its modern form as early as 1678–80.

As noted above, this use of W *ôl* is comparatively early and may have been used in relation to ways or trackways which could in more recent times be implied by the common *heol* 'road, way, street'. Early examples occur with the names of domesticated animals to indicate well-worn or well-marked tracks. In the *Liber Landavensis*, and in the OW orthography of that compilation, there occurs *ol huch*, ModW *ôl hwch* 'sow's track' and *oligabr, ol ygabr* ModW *ôl y(r) afr* 'goat's track' to denote tracks which ran along district or parish boundaries and which are reminiscent of the 'stock tracks' along township boundaries in north Lancashire to which Mary Atkin has drawn attention. In this example it is highly possible that the reference is to a path or track on the boundary line between the parish of Welsh St Donats (Llanddunwyd) and the detached portion of Llancarfan.

Similar examples of the name are Olmarch in the parish of

Llanrheithan, Pembs., and the names of four farms between Tregaron and Lampeter (Llanbedr Pont Steffan) in Ceredigion, Olmarch Fawr, Olmarch Ganol, Olmarch Isaf and Olmarch Cefn-y-coed.

Homri

This farm stands to the north of the village of St Nicholas on a spur overlooking the Ely river and opposite the village of Peterston super Ely. In its present form the name is unique, but it is known to toponymists as one of the three relatively certain examples of names in the vicinity of Cardiff which contain elements which indicate probable settlements of Scandinavian origin.

The three contain the ON element *bȳ*, *býr* 'homestead, farmstead' which survives as the suffix –*by* in modern forms, the earliest recorded forms of the name Homri being *Horneby* 1382–3, 1540, *Hornby* 14c., *Horne Bye* 1572, *Hornebye (Wood)* 1591–5 etc., with a first element generally regarded as being the masculine Scandinavian personal-name **Horni*, although the ON *horn* 'a horn-shaped piece of land', that is, a projecting tongue of land, or spur, remains a possibility having regard to the location above the Ely valley. It can be compared with two places called Hornby in north Yorkshire, the one near Great Smeaton and the other near Hackforth, and the well-known Lancashire Hornby, all of which can be similarly interpreted. The other two names with the suffix -*by* in the Cardiff area are the street-name Womanby Street which preserves a form of the original *Hundemanby c.* 1280 < ON *hundamaðr* 'houndsman, dog-keeper' + *by*, and Lamby on the Rhymni estuary to the east of Cardiff which is *Langby* 1401 < ON *langr* 'long' + *by* now preserved in another street name, Lamby Way.

Despite the suffix -*by* in these names, it is not possible to define exactly what kind of settlement is indicated by its presence, no more than with other names which are regarded as being of Scandinavian provenance in Wales. The main reason for this is that they are not evidenced in sources which are early enough to provide such accuracy. To take the three examples quoted above, they are first recorded *c*. 1280, 1382–3 and 1401 respectively, all comparatively late forms of names of settlements that could have been established centuries previously. The main thrust of the Viking attacks on Wales came by way of Ireland and was related to their activities in the Irish

Sea during a period which historians now recognise as having had phases of contact, or a pattern of sub-periods commencing with the punitive raids of the ninth century and extending to the somewhat less destructive incursions of the eleventh. There had been a further influx of non-Welsh influences up to the time of the first recorded appearance of the names, beginning with the Anglo-Norman presence which closely followed the Scandinavian, so that the investigator is faced with problems of interpretation which have to take note of the further semantic development of an element like ON *by*, even its absorption into Middle English. As the late Professor Kenneth Cameron reminded us, it is 'a living word by the early ME period'.

The exact nature of the settlement at Homri cannot be precisely determined. Its location is of interest, a few miles inland up the Ely valley, which may encourage speculation that it could have been an agricultural settlement rather than a trading station. That it became a substantial holding later is borne out by entries in the land tax assessments of the late eighteenth century where reference is made to two taxable properties, *both Hombries* 1784–5, 1791, *Hombreys* 1792, implying its division into two portions. These forms also suggest that the stages in the evolution of the modern form of the name were *Hornby* > *Honbry, Honbri*, by metathesis –*rnb*– > –*nbr*–, then *Hombri(e), Hombrey* by the well-attested colloquial change of –*nb*– > –*mb*–, and finally *Homry* (1816–38), *Homrey* (1824), *Homri* (1833), the –*b*– becoming a lost consonant as in the words *tomb, comb* etc. Loss of the initial aspirate in local pronunciation and analogy with the biblical personal name together may have suggested to the local incumbent the form *Omri*, which appears in the parish register, 1836.

Lavernock : Larnog

This name seems to have caused some unease among generations of speculators concerning its meaning. Explanations range from a claim that it is an indeterminate Old Scandinavian form to a far more reasonable suggestion that it may be a form of the W adjective *llywernog* in a substantive sense, this being a compound of W *llywern* 'fox' and the adjectival suffix *–og* to give the sense 'a place frequented by foxes'. This is reinforced by the similarly named stream Llywernog, which runs down its little valley near Ponterwyd, Cards.

This theory seems to have been suggested in ignorance of the attested dated forms of the name and the local pronunciation *Larnog*, or something like *Larnock* by non-Welsh inhabitants. This is confirmed by forms such as *Lannock* 1537, *Laurnock* 1584, *Larnott* 1596–1600 (*tt = cc*) and *Larnoc c.* 1678.

In addition, the majority of attested forms strongly imply that *laver–* is the original form of the first element: *Lawernak* 13c., *Lavernock* 1425–6, 1533–8, 1609, 1635, *Lavernorke* 1529, *Laverneocke* 1558, *Lavernoke* 1563 etc. Similarly, it is difficult not to see the common English *knock* 'hill, hillock', OE *cnocc*, as the second element, and whatever the meaning of the element *laver–*, the *knock* suits the location as most of the land that constitutes the small parish lies on the slopes of a prominent hillock, much of which has now been eroded by the tidal action of the Severn Sea on its southern side.

It looks like an English name, and it may be profitable to compare it with names of places on the opposite side of the Bristol Channel, particularly Laverstock, Laverstoke and Laverton in Somerset, a region which has supplied many immigrants into the Vale of Glamorgan over the years. The accepted view of *laver–* in these names is that it is a ME form of OE *lāwerce* 'skylark' which gives *lark* in ModE, and if this is the first element of Lavernock, it would have the meaning of 'a hill frequented by larks'.

It is an attractive and admissible suggestion for the association with skylarks is supported in the observations of a notable Welsh literary figure, Saunders Lewis, who, in his short Welsh poem entitled 'Lavernock', includes the following brief descriptive couplet

> Gwaun a môr, cân ehedydd
> Yn esgyn trwy libart y gwynt.
>
> Moorland and sea, the song of the lark
> ascending in the liberty of the wind

Lecwydd

The prevailing interpretation of the name of this village between Cardiff and Penarth which appears consistently as Leckwith is that it is a form of W *llechwedd* 'slope' as the parish and the village stand on a ridge which rises from the west bank of the river Ely (Elái). Samuel Lewis, in his *Topographical Dictionary* (1843) gives voice to this claim: 'Leckwith (Llechwedd or Llechwydd) ... the name signifies "the slope of the cliff" ... ' and has been followed by many.

If there is any truth in this, it must be assumed that the medial provection of the Welsh *–ch–* to *–c–*, (*–ck–*), had occurred by the middle of the twelfth century for there is no vestige of the spirant *–ch–* in the collected forms of the name of which there are more than is usual available. It first appears as *Lechwyth* 1578 by Rice Merrick of Cottrel which looks very much like a contrived form in order to give it meaning.

The attested early forms are too numerous to be listed here, but the earliest is *Leocwth(a)* 1153–83, followed by *Lecwithe* 1179, *Lequid* 1184–5, *Lecquid* 1233, *Lekwith* 1306-7 etc.

The suggestion has been made that the name represents the name of an early Christian saint which would appear in ModW as *Helygwydd*. Such a name appears in the *Liber Landavensis* as *Helicguid*, *Elecuid*, and in the name of a holy well in Gwent, *fynnaun elichguid* in the same source, which is probably compounded of W *helig*, *helyg* 'willows' (Latin *salix*) + the collective noun *gwŷdd* 'trees, wood', and however uncommon such a personal name may appear both these elements are attested as elements in OW personal names.

Whatever the exact significance of *Helygwydd* as a personal name its existence cannot be denied. Given this fact some changes to its original form would have to be accepted if it is to be regarded as the basis of the place-name *Lecwydd*, the first being that the provection of medial –g– to –c– between a vowel and –w– would have had to occur as early as the twelfth century. Examples of such a change are not plentiful but they do exist, as in *Wicwer, Wicwair, Wicwar* (*Villata de Wickwere* 1335), colloquial forms of Wigfair (Denb.), or the masculine personal name Tecwyn from Tegwyn. Also, the loss of the first syllable *(H)e–* to give the form *Lecwydd* can be compared with a similar loss in old Welsh personal names like *Leri* < *Eleri*; *Liwlod, Lywlod, Lywlat* < *Eliwlat, Elyflad*; *Liddon* < *Eliddon* (see p. 119) and the colloquial *Lai* for the river name Elái (Ely).

If this interpretation is valid, we have here an example to set among those where the name of the saint to whom the parish church is originally consecrated, like Ceidio, Baglan, Gwytherin, Llywes, has become the name of a village and parish without the usual qualifying element like *llan, eglwys* or *merthyr*.

Litchard

Over two centuries ago this name of the populous northern district of Bridgend was that of a single homestead, *Lydiat farm c.* 1798, *Lydiad farm* 1846, but more surprisingly, bearing in mind the error-ridden reputation of early map-forms, given correctly on the first edition of the Ordnance Survey one-inch map in 1833 as *Llidiart*. There can be little doubt about the validity of this form, corroborated as it is by a number of dated forms.

The first of these occurs in a lease of 1452 of what are called three tofts in the Great Park of Coety, one of which is *Lydeard*. The term toft is interesting because it is a borrowing into English from Old Danish and had the specialised meaning of 'a building plot, a curtilage'. The best evidenced meaning in English is 'the plot of ground on which a dwelling stands'. This of course has no connection with W *llidiart*, an appellative meaning 'gate, wicket; barrier', itself a borrowing from OE *hlidgeat* 'a swing-gate', especially one to prevent cattle straying from pasture on to arable land or to another location, possibly woodland. The form Litchard has evolved under English influence.

The naming sequence appears to be that an original plot of land was named because of its proximity to a point of entry or exit between two such recognised areas or on the boundaries of which there was a prominent barrier. This is confirmed by the fact that given the early designation of *toft*, it appears significant that many collected forms of the name are compounded with W *cae* 'field' and the W definite article, thus *Cay yr lidiat* 1598, *Kar y llyddiatt* 1635, *Kae yr llydiard* 1636, *Caer Lydiart* 1638, *Caerlidiart* 1639–40, *Kae'r Llyddiart* 1720, *Kaer lidiad* 1721, 1726, *Cae'r lidiad* 1786.

Furthermore, this is in the parish of Coety where a number of places are called Pen-coed, Tor-coed, Coed-y-gaer, Coedypebyll, Coedymwstwr and several others where the common element is W *coed* 'wood, forest', the old member lordship of Coety having been held of the Norman lord of Glamorgan by a particular form of tenure known as serjeanty which gave the lord liberty to hunt in the lordship's woodlands, a feature of its topography thus reflected in local nomenclature. The significance of the presence of a barrier of some sort to control the movement of livestock is also emphasised.

As an attested element in Welsh place-names *llidiart* is reasonably well-evidenced. The plural form Llidiardau occurs near Craigcefnparc in the vicinity of Clydach, Cwmtawe and shown on the 1884 Ordnance Survey six-inch map, this being *Llydiad y park* in 1751, *Llydiard-y-Park* or *the Mountain Gate* 1827, its proximity to the park, W *parc*, being significant, again as a means of controlling straying animals. Nearby stands Y Fagwyr farm, this being W *magwyr* 'wall' which may well occur in *Llidiart y fagwyr* in an undated document (possibly of the eighteenth century) relating to the same location. Either of these could be the precursor of the present Llidiardau, the local pronunciation of which, according to the late Professor Henry Lewis, was the singular Llidiate.

Llanffa

Originally the name of a mesne manor of the lordship of Ogmore (Ogwr), the present hamlet of Llanffa and the nearby Llanffa Court owe their name in all probability to the location of the site of a pre-Norman chapel, now no more than a rectangular mound in a field south of the Court. This was the *capella Ugemor de Lanfey* 1141 which never attained the status of a parish church. The prevailing form of the name on maps is Llampha, but this is merely the result of the outmoded fashion of representing the Welsh *ff* as *ph*, although it is recorded as early as the seventeenth century: *Lanphe* 1620, *Lanphey* 1631, 1649, *Lanphey* 1723 etc. for *ll.fai c.* 1566, *Llanffe* 1594, *Lanffee* 1596–1600, *Llanffay* 1600.

However, the Welsh form with the digraph *ff* is itself a colloquial form which evolved out of a stressed pronunciation of an original W *f* (= *v*) and may be the result of anglicising influences, cf. the pronunciation of the river name Taf as Taff, Llandaf as Llandaff etc. The earliest recorded form is *lan tiuei* in the *Liber Landavensis* which would be Llandyfei, later Llandyfai in modern orthography and which is W *llan* + the saint's name Tyfai (composed of the Welsh honorific prefix *ty* + the old pers.n. *Mai* > *Tyfai*), this being well documented from the twelfth to the fifteenth century, *Landefei* 1139–48, *c.*1260, post 1281, *Landevai* 13c., *Landefey* 1317, 1328–9, 1350, *Lantefey* 1332 etc. During this period there must have been a tendency to lose the unaccented first syllable of the pers.n. as is shown in the 1141 form *Lanfey* noted above and which occurred in the similarly composed name of Lamphey (Lampha), Pembs., (*Lantefei* 12c.) whereas in a similarly stressed pers.n. in the name Llandygái (Llandegái), Caerns., this has not occurred.

Another form of the name with the additional element W *sant* appended is Llandyfeisant, near Llandeilo, Carms., where the honorific prefix has survived in a different stress pattern. More problematical is Llanfoy in Herefordshire which is *Lann tiuoi, lann timoi* in *Liber Landavensis* with the hypocoristic saint's name Tyfoy or Tyfwy based on the personal name Moe or Mwy which has been claimed to be a variant form of the Mai in Llandyfai (Llanffa).

Further, if the first element of Martletwy, Pembs., is an anglicised form of the W *merthyr*, as has been claimed by Dr B.G. Charles in his study of Pembrokeshire place-names, it is virtually certain that the second element in such a name is that of a saint. In this case it may have been Tywai, a possible by-form of Tyfai, the interchange of medial *f* and *w* being attested elsewhere in Welsh, as in *cafod/cawod*.

Llanfihangel-y-pwll

This is the form of the name of the village of Michaelston le Pit, near Dinas Powys, which was used by the Welsh-speaking population of the Vale of Glamorgan since at least the sixteenth century. It is recorded as *ll.fihangel or pwll c.* 1566 in an early list of Welsh parishes, with the variant form *ll.V'el y pwll* in a later version of 1590.

The English name is evidenced much earlier as *Michelstowe* 1291, *Michelestowe* 13c., *Mighellesstouwe*, *Mighellstowe* 1307 etc. from which it is clear that the second element was originally *stow(e)*, OE *stōw* 'holy place, place of assembly' and not the *-ton* by which it was supplanted in the later forms (sometimes in the modern form *town*), again from the sixteenth century onwards: *Michaelstown* 1535, *Mychelston* 1559, *Mighelstowne* 1565, *Mighaelston* 1590 etc. It originates as the name of the church, an Anglo-Norman dedication to Michael the Archangel, first recorded in the Norwich Taxation of 1254 as *Sancti Michaelis de Renny*, the feudal appellation being that of the de Reigny family who held the manor in the thirteenth and early fourteenth centuries.

Although the evidence is not available, it seems likely that the W forms in *llan–* first come into circulation to render the forms in *–stow(e)* rather than those in *–ton*. The substitution of English elements appears to have occurred in the first half of the sixteenth century, this being consistent with what occurred to other names of a similar kind, notably Tythegston (Llandudwg) which was *Tethegstowe* 1258 but *Tythegston* from 1513 onwards and a Welsh dedication in the first place. The change may have been the consequence of a simple sound analogy between *–stow(e)* and *–ton* when preceded by a genitival *s* (*–ston*). Also the widespread occurrence of names in *–ton* in the area, both ecclesiastical and secular, may have been an added consideration.

The additional elements *le pit* in the later English form and *y pwll* in the Welsh form of the name are obvious distinguishing elements in view of the proximity of Michaelston super Ely or Llanfihangel ar Elái and Michaelston y fedw or Llanfihangel y fedw. It is also noticeable that these elements begin to appear in documents about the mid sixteenth century in the case of the Welsh form. *Mighelston in le Pitt* appears in 1567, *Michaelston the Pitt* 1596, *Mychelston le Pytte* 1603 etc., with English *pit* and Welsh *pwll* being virtually synonymous in this context and descriptive of the location of the site of the parish church and present village. They lie in a wide basin-like depression and the use of OE *pytt* in the sense 'a natural hollow' is normal, particularly in the English West Country. Welsh *pwll* has a similar connotation in such a context although it can be equated with English *pit* in other senses (cf. *pwll glo* 'coal pit', *pwll clai* 'clay pit, *pwll gro* 'gravel pit') and has also the watery sense of 'a pool'.

In view of the close proximity of the dates of the forms in which these elements make their appearance it is hardly possible to ascertain which is the earlier, *pit* or *pwll*. One recorded form, *Michelston le Pole* 1563 caused considerable speculation at one time. It certainly looks like a 'one off', rather than an affixed manorial family name, in this case that of the dukes of Suffolk who possessed the manor before 1493, as was once claimed. The form never gained currency, as might have been expected if that were true, and it may well be the common ME *pole*, OE *pōl* 'pool' used because of its resemblance to W *pwll* in that surviving form of 1563.

The use of the French definite article *le* which has survived to modern times in the English form is first seen in 1563 but its complete absence from all the earlier forms cannot sustain a claim of Anglo-Norman provenance. It has been shown that many existing E place-names which exhibit a similar element in their m. or f. forms seem to have acquired it later, in the eighteenth century in particular, among them Stratford le Bow, Holton le Clay, Newton le Willows etc. in England in which *le* has acquired the function of a preposition having the sense 'in', 'on' or 'by' which developed as a formula. Both appear in the form *Mighelston in le Pitt* 1567.

Llanilltud Fawr : Llantwit Major

The attempt to resurrect the spectre of a 'twit' in the so-called 'corrupt' English form of the name of Llanilltud Fawr in the Vale of Glamorgan continues in popular writing. That the form Llantwit is not corrupt is demonstrably evident and that to believe that the English slang term *twit*, which is not evidenced in the OED until 1934, is part of a name recorded as *Landiltuit* as early as 1106 is absurd.

What we have, after the W *llan*, as the second element of both forms of the place-name, Llanilltud and Llantwit, are two forms of the name of the same saint to whom the church is dedicated, indeed, the founder of the original early sixth-century monastic cell. One form is W Illtud, the basic form being probably Elltud (as in Llanelltud near Dogellau, Merionethshire, a solitary dedication to the saint in north Wales), *Laniltuth* 1254, *Lan Iltut* 1295–6 etc. The other form is probably a Goidelic, or Irish genitival derivation of the same Celtic root-form, Illtuaith, which is seen in the 1106 form *Landiltuit* noted above, *Landhiltwit* 1184–5, *Landiltwyt* 1256 etc., later to appear as *Laniltwit, Laniltwyt, Lanyltwyt* etc. from the twelfth to the sixteenth centuries. It is also relevant to note in relation to this form of the personal-name the earlier forms of another Illtud foundation at Ilston in Gower which was *Ilewitestoun* 1319, *Ilewitteston* 1396, *ecclesia de Sancti Iltuti de Illiston* 1490.

By 1316 the form *Lanntwyt* is recorded, *Lantuyt* 1361, *Lantwytt* 1431 etc., where the second syllable is lost in oral transmission. This second form is what survives as the modern Llantwit. In Welsh orthography it would have been Llantwyd and this is probably on record in the forms *Lantwyd(e)* 1536, 1548, 1565, *Lantwid* 1695, *Llantoid c.* 1558, for it is the form recorded for an Illtud dedication in Pembs. which later became Llantwd, subsequently anglicised as Llantood.

What has caused many to regard Llantwit as an 'English' form is the addition of *major* to correspond with W *mawr* 'great' in Llanilltud Fawr. This is obviously an ecclesiastical and documentary addition which bestows a higher ranking on this ancient foundation over the three others of the same name in Glamorgan, Llantwit juxta Neath (Llanilltud Nedd or Llanilltud Fach), Llantwit Fardre (Llanilltud Faerdref), Ilston (Llanilltud Gŵyr), and Llanilltud or Capel Illtud in Breconshire. Llantwit Major is also probably the prototype of the other forms in Glamorgan, bar Ilston.

Llantriddyd

This a small village in the Vale of Glamorgan which is noted for the substantial ruins of Llantrithyd Place to the south-west of the churchyard, the house of a branch of the Basset family of Old Beaupre in the early sixteenth century.

It was the resourceful Edward Williams (Iolo Morganwg) who set the fashion for the belief that the original religious settlement here was founded by a *Treiddyd Sant* from Llanilltud Fawr, thus giving his name to what appears in another of Iolo's notes as *Llan Treuddyd*. This was accepted by many, including Dafydd Morganwg in his *Hanes Morganwg* (1874), and later an eminent ecclesiastical scholar widened the choice of a founding 'saint' by suggesting that it was the female Trynihid (Trinihid), the wife of the celebrated Illtud, *feminarum castissima*, according to hagiographical tradition.

Apart from the difficulty of accepting Trynihid as the possible second element of the name Llantriddyd on phonological grounds, the suggestion is immediately suspect in the light of the simple and well-known fact that in Welsh the initial consonant of the name of a saint, be it male or female, is lenited when it follows the feminine noun *llan*, as in Llandeilo (Teilo), Llandudno (Tudno), Llandwrog (Twrog) etc., but here we have Llantriddyd, not Llandriddyd.

While this may raise doubts about the second element in the name, it also suggests that the original first element may not have been *llan*. The early attested forms, *Lanririth* 1254, *Lanririd c.* 1262, *Lanryred* 13c. etc., show no trace of *t–* or *d–* as the initial consonant of the second element, and when we come to the forms of the name which originate from Welsh sources from the sixteenth century on, what occurs is *Nantririd c.* 1545, *Nant Ririd* 1569 etc., becoming *Nant Triryd* 1550–1600, 1606–20 under later influences.

What such evidence implies strongly is that the second element in the name is the old Welsh masculine personal name Rhirid, becoming *–rhiddid*, *–rhiddyd* by assimilation, together with the fact that it was the W n.f. *nant* 'valley, glen', later developing semantically as 'stream, rivulet, brook', and not *llan*, which is the original first element. The medial *–t–* of the present form of the name which has been taken erroneously to be the initial consonant of the second element is, rather, the vestigial terminal *–t* of the original first element *nant*.

It is known that the substitution of *llan* for an original *nant* occurred quite early in Welsh place-names, particularly where the second element is a personal name and where the locality had strong early Christian associations. Well-known examples are Llancarfan, Llantarnam (Nant Teyrnon), and Llanthony (Llanddewi Nant Hoddni). Furthermore, an indication that the true significance of the second element in Llantriddyd had long been forgotten locally (Rhirid here being an example of the use of a personal name as a stream name, an unusual but not uncommon occurrence in Wales) is that the stream which runs through the village is not now called by its original name but is referred to by the tautologous form, Nant Llantriddyd.

As is noted in the discussion of Llwyneliddon (p. 120), evidence exists to show that *llan* and W *llwyn* 'bush, grove' were also interchangeable as first elements of some place-names, forms of a basic *Llanliddan* being evidenced for Llwyneliddon in the sixteenth century. Conversely, *Lloyntrithed* 1596, *Lloynrithed, Llwyn Ridit* 1657–60, *Loynrithyd* 1684–94 occur for Llantriddyd, possibly by analogy with *Llwyneliddon*, but the form did not endure.

Finally, the anglicised spelling Llantrithyd, where the 'soft' English *th* is used for the Welsh *dd*, as in Aberthaw, Llanbethery, Llanblethian, is not to be encouraged.

Llanyrnewydd (Llanenewyr)

This is the unusual and seemingly almost meaningless name of a church which was once a chapel of ease of Llanrhidian, Gower, until the grant of parochial status in 1924. As it stands Llanyrnewydd appears to be composed of three Welsh elements, namely the W n.f. *llan* + the full form of the Welsh definite article *yr* + the W adjective *newydd* 'new', a combination which hardly suggests a recognisable dedicatee or founder's personal name, certainly not what is suggested by the present official designation of St Gwynnour's, Llanyrnewydd. The form Gwynnour appears to be an anglicised form of a Welsh name Gwynnwr and it was A.W. Wade-Evans in his *Parochiale Wallicanum* (1911) who recorded *Llanrhidian Chapel or Llangwynner*, and the patron saint as Gwynnwr, based on Browne Willis's *Parochiale Anglicanum* (1733) where the name of the church appears as *Llanweynour*. This attribution may represent an attempt to establish some relationship with Llangynnwr, Carms., whilst another comparison which has been made is with Llanwnnwr in the parish of Llanwnda, Pembs. Attempts have also been made to locate the *lann conuur* or *cella conguri* of the *Liber Landavensis* in the area of Llanyrnewydd, but with no satisfactory conclusion.

The bulk of the evidence of the collected forms of the name does suggest a personal name as the second element but it is not such as to support or confirm the veracity of the forms noted above, as shown by the following attested forms: *Lanyynewis* 1499 (*s = r*), *ll.ininewyr* c. 1566 (leg. *ll.inewyr*), *Llangnewyr* 1584 (leg. *Llanynewyr*), *Llanynewer* 1587, *the Chapel Kae-y-newyn* 1598–1602, *Llannyenwere* 1610, *Lanridian o. Langwire* 1666, *Llanynewir Chappell* 1677, *Llan y newer chap* 1697, *Llanynewn Chapel* 1741–2, *Llan y new yr, Llanynewir* 1764.

The obvious choice is *Ynewyr, Ynewir*. This was noticed by

Egerton Phillimore, who also saw the resemblance to a saint's name preserved in three examples of a Breton place-name in the département of Finistère, namely Plounéour-Lanvern near Pont l'Abbé, Plounéour-Trez near Lesneven and Plounéour-Menez near Morlaix, names which are prefixed by Breton *plou*, W *plwyf* 'parish', the saint's name being *Eneour*, a name recorded in the cartulary of Landevennec in the phrase *plebs Sancti Eneguorii*. This suggests a parallel Welsh form *Enewyr* which became *Ynewyr* in a compound after *llan*, the strong accent on the second syllable causing the unaccented initial *e–* of Enéwyr in the four-syllable compound *Llanenewyr* (the probable original form of the name) to become indistinct, giving *y–*, and Ynewyr, hence Llanynewyr.

The immediate possibility thus presented to native Welsh speakers at a time when all knowledge of the significance of the second element had been lost, and in view of its accentuation, was to assume that the medial unaccented *–y–* was the vocalic form of the W definite article (Llan-y-newyr) carried a stage further in the nineteenth century by the restoration of its full form *yr*. In the meantime, popular etymology, probably based on a form of sound analogy, seems to have accounted for the fact that the still meaningless final element *–newyr* was interpreted as the W adjective *newydd*, thus producing the form Llanyrnewydd, also possibly aided by a colloquial change of final *–r* to *–dd*, a change not as commonly attested as that which occurs by assimilation in the combination *–n–r* > *–n–n* which would give *–newyn* as attested by the 1598–1602 and 1741–2 forms cited above, and further aided by the existence of the common Welsh word *newyn* 'famine'.

That a parallel Welsh personal name with the Breton *Eneour* (probably earlier *Enever*) is on record was demonstrated by Sir Ifor Williams in his consideration of the inscription ENEVIRI on a seventh- to ninth-century stone which once stood in Tregaron churchyard, Cards. Sir Ifor regarded it as a Welsh name, *Enevir*, with the Latin genitive ending, which would have developed later as *Enewyr*. He took it back to a British antecedent *Anauo-rix*, compounded of a word which survived as *anaw* in MW meaning 'beauty, wealth' + *rix* 'king', a derivation which also fits the Breton *Enever*.

Whether this provides a conclusive argument for regarding Enewyr as a Breton 'saint', as has been claimed, thereby initiating an attractive theory of settlement, is open to question.

Llawennant

Near the ALCOA works by the river Llan in the vicinity of Waunarlwydd and Mynydd-bach-y-glo, Swansea, stand the ruins of the old farmhouse of Ystrad Isaf 'the lower Ystrad', the partner of Ystrad Uchaf 'the higher Ystrad' which once stood to the east, near Fforestfach, its name now preserved in that of the present Ystrad Road.

Ystrad here refers to the low-lying land between the road from Swansea to Loughor (Casllwchwr) and a small stream which runs into the Llan nearby once called Nant Cwm-bach. It is a Welsh word of Celtic origin which, possibly because of the known Latin renderings of names like Ystrad Fflur as Strata Florida, Cards., or Ystrad Marchell as Strata Marcella, Mont., tended at one time to be regarded as derived from Latin *strata* (a form of the verb *sterno* 'to stretch out, spread out') in the sense 'a way, a paved way'. As Sir Ifor Williams has shown, this is not the case. W *ystrad* is cognate with Latin *strata* but has the OIr parallel form *srath* 'valley bottom, flat place' especially by a river or lake, and a rare OCo **stras* 'flat or shallow valley, sward, meadow', and as these forms show it was originally monosyllabic. It is found as *Strat* in the *Liber Landavensis* (*Strat Elei* for Ystrad Elái to the south-west of Cardiff). However, in common with some Welsh words which have an initial *s–* + another consonant (including some borrowings from Latin) a prosthetic *y–* developed before the consonant group with a subsequent accent shift on to the prosthetic vowel as in W *ẏsgol* < Latin *scāla* 'ladder' or *schola* 'school', W *ẏsbryd* 'spirit' < Latin *spiritus*, W *ẏsgub* 'sheaf' < Latin *scōpa* etc., to give W *ẏstrad*.

Normally, in Welsh place-names, *ystrad* occurs with either the name of a river or stream as in Ystrad Tywi, Ystradfellte, Ystradfflur, or a personal name as in Ystrad Marchell, Ystradyfodwg, Ystradowen. In the case of the Swansea Ystrad its full name contained the earlier name of the brook Nant Cwm-bach, namely the Llawennant. It is recorded as *Ystrad Llewenant, Ystrad Lewenant* 1600, *Ystrad llewnant* 1616, *Ystrad lawenna* 1650, *Ystradelawena* 1699, *Ystrad Llewennant* 1701, 1704, *Ustraed* 1729, *Ystrad llawrnant* 1764, *Ystradlawennant* 1764. The surrounding landscape was wet and boggy so that associated with the stream name there existed *Cors Llawennant*, with W *cors* 'swamp, bog, marsh', the final consonant group being lost in common parlance to give a form like *Cors Llawenna*, recorded as *Gorse Llawenna* 1650 and *Cors Lewenau* on the first edition of the Ordnance Survey one-inch map, 1833.

The prime authority on Welsh river names, the late R.J. Thomas, like Sir Ifor Williams before him, was inclined to see in the name the W adjective *llawen* 'merry, jovial, cheerful, happy' and notes a number of Welsh river names which are so named and others which are derived forms + W *nant* 'stream' (earlier 'glen. valley') but there may be an alternative indirect way in which this meaning could have been acquired in this case.

Most boroughs of Anglo-Norman origin were alien plantations in the Welsh environment when first established. Burgesses were largely of non-Welsh origin and local place- and street names predominantly English. It was not until the later medieval period that the native Welsh began to play a part in borough affairs with a consequent gradual reassertion of Welsh influence in local nomenclature, even a measure of renaming in some instances.

The area in which Ystrad Uchaf and Ystrad Isaf stood was on the edge of the borough of Swansea in the Middle Ages, near the Portmead which survives as the name of a suburban area (with *port* here in its earlier sense of 'market town, borough' + *mead*, as in *meadow*). The stream Llawennant once formed the boundary between the Portmead and the land on its western side, Crow Wood. The well-evidenced ME term for a boundary or border is *meare*, *mere*, OE *gemǣre*, and since streams and brooks formed natural boundaries some have been designated as such and named by using *mere* as a qualifying element with OE *brok* 'brook', this becoming a place-name in some cases like Meerbrook in Staffordshire or Meersbrook in Derbyshire. B.G. Charles records a *Meare Brooke* 1612 in Monmouthshire. The use of ME *brok* in one Swansea borough boundary name is evidenced, with a different qualifying element, in *Burlakesbrok* 1153–84, traces of which are retained in the modern Cwmbwrla (pp. 57–8). Could the Llawennant have been named a *merebrok* at one time, and could the same thing have happened as occurred in the case of a *Merebroc* c.1200 in Worcestershire where it was later misunderstood as *merry-brook*, surviving as the place-name Merry Brook? If so, the use of W *llawen* 'merry, cheerful' etc. + *nant* as a literal Welsh translation can be understood in a period of reimposition of Welsh local nomenclature. No evidence of such a change is available, but it is possible.

Llechwen

The present Llechwen Hall is a well-known hostelry in the parish of Llanfabon, the core of the property being the house which was formerly the home of a branch of the Thomas family of Llanbradach Fawr. Some features of the building can be dated to the seventeenth century but it is predominantly an eighteenth-century structure.

Its original name was Llechwenlydan with the W adj. *llydan* 'wide, broad' appended and initially lenited after a feminine noun but is now lost. Although this name possibly predated the house as being descriptive of the proximity of its location, no forms earlier than the eighteenth century have been found but, making due allowance for the orthographic peculiarities of non-Welsh scribes, those that are recorded would seem to corroborate the original form, *Lechwan Ledan* 1783, *Llechwanlodan* 1785, *Llechwanllydan* 1826, *Leachwenleodan* 1827, *Leckwanllydan* 1829, *Llechwan-llydan* on the 1833 Ordnance Survey one-inch map and a reference to *Llechwen* farm 1888 and later.

Llechwen consists of two elements, the first of which is the W n.f. *llech* 'stone, rock, flat stone' also 'slate' (particularly in its diminutive form *llechen*, *llechan* in north Wales), and it occurs as an element in names like (Y)Benllech, Anglesey, Harlech (Har(dd)lech), Merioneth, in the sense of a large rock or slab. The second element has been wrongly interpreted as the feminine form, *gwen*, of the W adjective *gwyn* 'white, light coloured' probably in a wider sense of 'pale, greyish white' in reference to the colour of the stone, the exact nature of which is now impossible to determine.

Evidence is adduced in GPC for *llechwen* and its variant *llechwan* as a colloquial form of the compound *llechfaen*, particularly in south-east Wales, in some respects a tautologous compound where the second element is W *maen* 'stone, rock'. The 'two tables of stone' which Moses received on Mount Sinai are *dwy lechfaen* in one passage in the Welsh Bible (Deut. 4, 13). The form *llechwen* results from the further Welsh colloquial change of the medial –*f*– to –*w*–, as in *twrf / twrw*, *cwrf / cwrw*, *cafod / cawod* etc. and the reduction of the diphthong –*ae*– to –*e*– and –*a*– > *llechwen*, *llechwan* in common parlance. A similar consonantal variation can be seen in the place-name Corwen, Merioneth, which was *Corfaen* originally, *Corvaen* 1254, 1291. Also north of Ynysybŵl in the parish of Llanwynno is the farm Y Llechwen which was *Llechfaen* originally.

In addition to designating some natural or man-made feature, *llechfaen* is also the Welsh term for 'a bakestone', that which was placed either over an open fire or in an oven on which to bake unleavened bread or the small round cakes known as 'Welsh cakes', buns or small loaves sometimes known as *bara llechfaen* or bakestone cake. Another term used for such cakes in Glamorgan particularly was *pic*, plural *picau*. The term used in the Swansea valley for Welsh cakes was *picau lap* (*lap* 'soft and wet' referring to the mixture of ingredients). It is interesting to note, therefore, that among the forms collected of the name Llechwen near Ynysybŵl there is an otherwise difficult reference to it as *Lach Wen y Pi Cairn* 1824, where *Pi Cairn* may well be an attempt to rationalise the plural form *picau* with reference to a stony cairn-like feature to give it some meaning, some garbled knowledge of the use of a *llechfaen* to bake *picau* being the underlying concept, for it seems unlikely that the name refers directly to a bakestone.

Another name found in Glamorgan which contains *llechfaen* as an element is that of a prominent hill above the Lluest Reservoir at the upper extremity of the Rhondda Fach valley, Bryn Llechwenddiddos. Also *a stone called llechwen dduddos* is documented in 1611 which is *y Llechwen ddidoes* 1628, 1638, *llechwen ddiddos* 1654, located (as far as can be judged from the sources in which the forms are found) in the vicinity of Y Gaer Fawr on the northern slopes of Mynydd y Gaer east of Briton Ferry. In addition, the Glamorgan poet Lewis Hopkin who resided at Hendre Ifan Goch, near Gilfach Goch in the parish of Llandyfodwg (Glynogwr) referred in correspondence with his son in 1768 to a Heol Llechwenddiddos 'in our neighbourhood'. GPC defines *diddos* as a W adjective 'watertight, weather-tight, sheltered' and as a n.f. 'shelter, place sheltered from the rain, rain-proof covering' and when compounded with W *carreg* 'stone, rock' as *carreg ddiddos* 'dripstone'. If *llechwenddiddos* refers to a natural feature it could be a projecting stone or slab which gives shelter or throws off water. If to a man-made feature it could conceivably refer to a supported slab or capstone, once known as a cromlech, such as those associated with ancient burials, and thus affording a rough shelter. However, no such structure is listed in the Royal Commission on Ancient and Historical Monuments' Glamorgan inventories in the vicinity of the locations mentioned above.

Lledglawdd

This is the name of an old farmstead on the western outskirts of Swansea and in the vicinity of Hendrefoilan and Craigybwldan (pp. 50–1). Lledglawdd is the form of the name which appears on modern maps but although it has no great significance as an establishment, as far as is known, its name has undergone more changes in its form over the years than many other similar homesteads. Indeed, there is some doubt as to the meaning of the name in its present form.

As it stands, its composite elements seem to be the W adj. *lled* 'half, part, partial' and the n.m. *clawdd* which, like the OE *dīc* 'ditch, trench' which acquired in ME the sense 'embankment, dike (dyke), wall', i.e. that which was thrown up when excavating a trench, also occurs in both senses in Welsh place-names, the verb based on the form being *clodd(io)* 'to dig, delve, excavate'. Had we here, therefore, a 'partial' or half-complete dyke or ditch at one time? It does not sound convincing, especially in the light of the earlier forms of the name.

The earliest of those forms evidenced to date is *Llechglawdd* 1652 which, with W *llech* 'stone, rock, slab' as the first element (cf. p. 114) makes it more likely that the second element *clawdd* here means 'wall, dyke' made of stone. Llechglawdd as the surviving name of a farm occurs near Llanboidy, Carms.

In 1631, however, the property and its land are recorded as a virtually meaningless *Llethclawdd, Tyr Cleth Clawdd*, and there seems to be only one way to account for this, namely that it is a well-evidenced scribal error where *ch-* in Llechglawdd has been read as *th-*, and where the *cl-* in *Cleth* probably appears because of a non-Welsh scribe's difficulty in pronouncing, and therefore spelling the fricative Welsh *-ll-*. Even so, the form Llethclawdd must have had some currency for by 1764 the name appears as *Llethrglawdd* in an attempt to give it a locational connotation, the intrusive *–r–* giving the impression that the first element is the W n.f. *llethr* 'slope, hillside, incline', recorded as late as 1852 which, in turn, was subjected to a form of cymricisation to produce the current Lledglawdd, still ambiguous in its meaning and having lost its original significance.

Llodre

At first glance, this appears to be a W plural form having the common colloquial ending *-e* for *-au*, as in *pethe* for *pethau* etc. in which case the primary form would be *llodrau*, a most unlikely candidate for use as a place-name element as it is a plural form of W *llawdr* 'trousers, breeches' but itself also used mainly in the singular sense. Because of this it comes as no surprise to find that subtle attempts from a relatively early period have been made to modify, indeed to change the form of some names which contain *llodre* in order to give them an air of respectability.

Possibly the earliest form attested is that which appears in OW orthography as *hen lotre elidon* in the *Liber Landavensis* (Hen Lodre Eliddon in modern orthography) as noted under St Lythans (see p. 119), a name which has been completely lost in common usage, the location also being unknown.

In the Sketty area of Swansea there occurs *Lodre Bryth* 1583, *Llodrybrith* 1764, where the W adj. *brith* 'speckled, mottled' is added as a descriptive element in another name which does not seem to be in circulation at the present time.

Llandremôr (accented firmly on the final long syllable), near Pontarddulais, took the form *Lladremor* before 1569, *Llodremor* 1584, and is *llodre* + the old Welsh personal-name Mor, but by 1764 *llodre* was superseded by *llandre* (as in the name Llandre, near Aberystwyth) possibly under the influence of the very common W *llan*, although the form *llandref* (i.e. *llan* + *tref*) is first evidenced in GPC in 1601.

In Llety Brongu, in the Llynfi Valley, the W n.m. *llety* 'temporary dwelling, lodging, abode' has replaced *llodre*, for the attested form in 1570 is *Llodre Brangye*, with *Llodre Brangig* 1548, and where R.J. Thomas sees another old Welsh personal name, Brangu, as the second element, its significance being that it could well be the *Brancu* which appears on a late ninth- or tenth-century stone found at Baglan.

The name *Gelli Lotre*, *Gelli Lotra* is also recorded at Llangeinwyr in 1833, and other examples are evidenced in the parishes of Llangennech and Pen-bre, Carms., including the possibly related adjectival form Llotrog near Penclawdd in Gower, which is *Laydrogg* 1688, *Llodrog* 1748.

It is more than likely that this element has a religious significance, to judge from its use in the majority of place-names in which it occurs. It is generally agreed that it is cognate with Ir *lathrach*, 'site or location of a house or church, or similar edifice', according to Sir Ifor Williams, but little more can be added at present.

Llwyn Eithin

Near the industrial estate of Fforest-fach, Swansea, there are a number of streets that have been named in Welsh after groves or thickets (W *llwyn* 'shrub, thicket, grove, copse') of trees and bushes of various kinds, Llwyn Derw, oak, Llwyn Celyn, holly, Llwyn Bedw, birch, Llwyn Helyg, hazel, and also Llwyn Eithin, with W *eithin* 'gorse, furze'. It may well be that it was this latter form which established the fashion, for it has a longer and more unusual pedigree than the others.

This area also, as has been noted elsewhere, was part of the outlying territory of the medieval borough of Swansea, hilly and wooded. The name of that portion which lay to the south of the river Llan was *Brynne canathan* 1583, *Brinkanathan* 1585, but *Bryn Clanathan* 1650, all these forms being recorded in documentary sources likely to have been written by non-Welsh scribes and containing W *bryn* 'hill' with a second element which is difficult to interpret as it stands.

However, in a will of 1562 there is a reference to what may well have been part of the same area but in the form *Pen lloyne Aythan*. In modern Welsh orthography this would be Pen Llwyn Aeddan, having W *pen* 'head, end of' + *llwyn* + the old personal name Aeddan. In view of the well-known fact that non-Welsh speakers and scribes had their difficulties with the Welsh fricative -*ll*- (as already noted in the discussion of the name Lledglawdd which also happens to be quite close to the area in question, p. 116) it could be that the second element in the form *Bryn Clanathan* 1650 and its forerunners is the scribe's attempt to represent Bryn Llwynaeddan (cf. the *lloyne Aythan* of 1562) with *pen* later taking the place of *bryn*.

The form Penllwynaeddan is attested by later forms: *Penllwynayd-van* 1741, *Penllynayddan* 1746, *Penllwynaythan* 1750–1, *Penllwyny-ddan* 1820, *Penllwyneithan* 1844, and in the last of these forms there is a probable clue to what occurred in common speech to produce the form shown on Emanuel Bowen's map of south Wales in 1729, *Llwyn Ithan*. In the local patois the Welsh diphthong -*ae*- becomes -

ei-, this in turn being reduced to a long *-ī-*, as in *eithin* becoming *īthin*. It is not difficult to accept an analogy here which 'restored' W *eithin*, pronounced *īthin*, to the written form of the name under the impression that this was its final element > *Penllwyneithin* 'the head (or end) of the gorse thicket'. The early stages of the process can be detected in an English scribe's entry in the hearth tax assessment of 1670 which is in the form *Pen lloyn Eythey*, with the final *–n* omitted.

There were farm units which bore the name at one time for the 1833 Ordnance Survey one-inch map shows that of *Pen-llwyn-eithin-fach* (sic) on the northern slope of Mynydd-bach-y-glo (the original *Bryn Clanathan*?) where factories now stand, and *Penllwyn-eithin* on the southern side, but they no longer exist.

Llwyneliddon : St Lythans

Not everyone is familiar with the Welsh form of the name of St Lythans, in the Vale of Glamorgan, but the collected documentary evidence which is available firmly supports its antiquity.

In one of the early charters in the *Liber Landavensis* which could well refer back to the seventh century, a reference is made to *ecclesia Elidon* 'the church of Eliddon', and this is the form which occurs consistently in ecclesiastical documentation from the twelfth century onwards, coupled with occasional references to *luin Elidon*, which is *llwyn Eliddon* in modern Welsh orthography, with W *llwyn* 'bush, thicket, grove, copse'.

In the same charter a reference is made to the location of *hen lotre Elidon* within the boundary of of the land of *ecclesia Elidon*, where the old Welsh word *llodre* (in modern orthography) 'site of a house or church or similar edifice' (p. 117) occurs after the W adj. *hen* 'old'.

It is probable that Eliddon in this context is an old W masculine personal name, particularly as reference is invariably made to *ecclesia Sancto Lythano* in documents up to the sixteenth century before the anglicised form St Lythans comes into regular use (although *St Lethian* is evidenced in 1400), based as it is on the Latinised form which was, in turn, an obviously truncated form of the Welsh Eliddon by loss of the unaccented first syllable (the English *th* here being 'soft' and used to represent the W *dd*, see p. 110).

However, it is also clear that by the sixteenth century the old form

Llwyneliddon was being used: *Lluen Lithan* 1536–9, *Llwyn Lyddon* 1550, *Lloinellithan* 1563 etc. with an assimilated form *Llwyn Ddyddan* by 1754–1805, but the reason for this remains unclear. The exact location of the *luin Elidon* of the early charter has never been determined, and it is hardly likely that the sixteenth century sees either a resurrection or a 'learned' colloquial restoration of the old charter form. It can only be assumed that this form continued in use in medieval times despite the scarcity of documented evidence for it during that period.

Further, there is considerable evidence in the early modern period for the existence of forms of the name with W *llan* as the first element: *Llan lidan* 1545–53, *ll(an) liddan* c. 1566, 1606, *Llan Leiddan* 1590–1 etc., and another assimilated form in Hywel Harris's *Welch Piety* in 1740, *Llanddiddan*, all of which could conceivably have come into existence under the influence of the English form St Lythans.

The form in *llwyn-*, therefore, has the support of the early charter evidence for the antiquity of its structure, but it is relevant to point out that although not plentiful, evidence does exist elsewhere to show that the elements *llwyn* and *llan* are interchangeable in common usage as is the case with Llwyngwarran and Llangwarran, Llwyngwathen and Llangwathen, both in Pembs., Llwynhywel and Llanhywel, Rads.

Llwynmilwas

The property which bears this name stands on the boundary between the parishes of Pen-tyrch and Llantrisant and is reasonably well attested, indeed better recorded in documentation than many similar properties in the area. With one important exception all the collected forms of the name confirm its present form if allowance is made, as it must, for scribal idiosyncracies.

It would appear that the full form was *Llwynymilwas*, the second element being preceded by the W def.art., which appears consistently in the seventeenth century: *Lloyn y Mylwas* 1630, 1666, *Lloyn(e) y Milwas* 1638, 1671, subsequently *Lloyn Mil(l)was* 1720–49 *Llwyn Milwas* 1752 etc. The exception happens to be the earliest form known to date, *Ll'n Mellwas* in the 1570 survey of the earl of Pembroke's lands in Glamorgan. The first element is clearly W *llwyn*, but the vowel variation in the forms of the second element, *milwas* and *melwas*, makes interpretation uncertain.

Melwas occurs as an old Welsh masculine personal name whose associations are inclined to be legendary, references to him being comparatively rare and indecisive. Such a figure is named in their works by the two major Welsh medieval poets, Dafydd ap Gwilym and Tudur Aled, and a person of that name is presented by Geoffrey of Monmouth as a man of royal lineage. This suggests the possibility that the two elements which constitute his name are W *mael* 'prince, lord', with the diphthong *-ae-* reduced to *e-* > *mel-* under the accent, and W *gwas* 'lad, youth', also 'servant, vassal', but no one of this name is known in relation to the Pen-tyrch–Llantrisant area, nor to the old commote of Meisgyn of which these parishes formed a part.

It may be relevant to note that Coed Melwas (with W *coed* 'wood, forest') occurs as a name in the parish of Coety, near Bridgend (*Kaer Melwas* in the will of Robert Gamage, 1570, with *cae* 'field, enclosure' + the affixed W def.art. *'r*, rather than *caer* 'fortress, stronghold'), but no personal connections are known in that respect, so that it seems prudent to look for composite elements which are common nouns in apposition. The possibility thus arises that we have here a form compounded of W *mêl* 'honey' and *gwas* 'servant' > *melwas*, one whose duties were to gather nectar for making honey, on the pattern of the known term for such a person, namely *melwr* < *mêl* + the agentive suffix *-wr*, defined in GPC as 'honey-merchant, provider or collector of honey' as in several recorded instances of the field name *Cae'r melwr*.

Otherwise, the tendency is for the form *milwas*, if this was the original form of the second element in the name, to be interpreted as a compound of W *mil* 'animal, beast' (as in the second element of Welsh words like *bwystfil*, *morfil*, *anghenfil* etc.) and *gwas*, that is, a person who looked after animals, possibly 'herdsman', in which case the *melwas* of 1570 has to be accepted as a variant form of *milwas*.

Llyswyrny : Lisworney

It is difficult today to imagine the erstwhile importance of the location of the small, pretty village of Lisworney, the anglicised form of Llyswyrny, in the centre of the Vale of Glamorgan. This is where the llys 'court, palace, habitation of a prince or nobleman' of one of the old cantrefi of Glamorgan once stood. The cantref was that which was known since the beginning of the fifteenth century at least as Cantref Gorfynydd, and by some, churchmen in particular, as Gronedd or Groneath, which became the name of a deanery in the see of Llandaf.

The variations in the collected forms of the name are too numerous to be listed here in their entirety, the earliest attested form of the name of the cantref being *guorinid, wurhinit* 12c in the *Liber Landavensis*, and of the location, *Liswrini* 1246, 1262, *Lyswreny* 1263, *Liswruni* post 1281 etc. It was the late Professor Melville Richards who showed clearly that the name of the region, or cantref, which is *guorinid* in OW orthography would probably be *Gwrinydd* in modern orthography and his reasonable and acceptable explanation of this form was that it contained an old personal name *Gwrin* + the well-attested Welsh territorial suffix *-ydd* (as in regional names like Eifionydd, Meirionnydd, Mefenydd etc.) 'the land or territory of Gwrin' (see p. 29).

The identity of Gwrin is not known, but that the personal name existed appears to be proved by its appearance, as the second element with W *llan*, in the name Llanwrin, Mont. Some further corroboration may be deduced from a hagiographical reference to a person named Gurai (Gwrai) (one of the sons of Glywys, ruler of Glywysing, which extended from the Tawe to the Usk) whose name is probably a hypocoristic form of Gwrin, as having inherited the land of *Gurinid* (Gwrinydd) on the death of his father.

The basic form which is the source of both Llyswyrny and Lisworney (a form which can be evidenced at least as early as the fifteenth century, *Leswornny* 1466, *Llysworney* 1495) is therefore *Llyswrinydd* 'the court of the land of Gwrin'. It will be noted that the initial consonant of the personal name is lenited after W *llys*, which is a masculine noun in modern Welsh but feminine in Middle Welsh.

The older preferred explanations are best disregarded. It is not *Llysyfronnydd* or *Llysbronnydd*, based on the W plural *bronnydd* 'hills', nor *Llys Ronwy* (which appears in fourteenth-century Welsh genealogies), based on the personal name *(Go)ronwy*, nor is there any connection with W *brwyn* 'reeds' as implied by John Leland's *Llesbroinith* 1536-9 and glossed in Latin *scirpetum* 'place of rushes'.

Machen

In a recent reproduction of the valuable volumes of Sir Joseph Bradney's *History of Monmouthshire* the author's tentative suggestion that the name of the parish and substantial village of Machen was a form of the saint's name *Meugan* has been allowed to stand without comment. It was once the name of an old Welsh lordship, a portion, perhaps, of Cyfoeth Meredydd, and was thus not necessarily likely to have an early Christian identity.

Further, the vowel in the first syllable of Machen in all its recorded forms is consistently -*a*- and not a diphthong, and although the medial -*ch*- caused problems for non-Welsh scribes in documentation, as the following forms show: *Mahhayn* 1102, *Maghein* 1439, *Machein* 1442, *Magh eyn* 1499–1500 etc., it is reasonably clear that such forms do not support the existence of a medial -*g*-.

The parish church is dedicated to Michael the Archangel, whereas the place-name contains the old Welsh personal name Cein, Cain, (which can be m. or f.) with the prefix *ma*- 'plain, open country', which can be seen in the common W *maes*, and in a lenited form in the much-used ending -*fa* in such words as *porfa* 'pasture', *morfa* 'moor' etc. and adapted in recent nomenclature to form names like Disgwylfa.

When *ma*- is used as a prefix it causes the spirant mutation in the initial letter of the following element. In the case of a person named Cynllaith, for instance, his open field or plain could be called Ma-chynllaith, now Machynlleth. Other similarly constructed names are Mathrafal, Mathafarn and possibly Mallwyd.

With the personal-name Cein, Cain, therefore, we have *Ma-chein*, *Ma-chain*, 'the open land of Cain' along the banks of the Rhymni, which became Machen, Machan in common parlance. A variant form is Mechain, in Powys, which is associated with the name of the river

Cain (possibly named after a person) but with a vowel mutation in the first syllable in that particular instance.

There is another Machen in Glamorgan, in the parish of Llangeinwyr (Llangeinor) in the Ogmore Valley, which was Eglwysgeinwyr at an earlier stage. The name here is compounded with W *cefn* 'ridge, back' in the form Cefnmachen, now the name borne by two farms but originally the name of the pronounced ridge of open land between the rivers Ogwr and Garw on which the church of Llangeinwyr stands—*Kevenmahhaj* (sic) *inter aquam de Vgkemor et Garwe* 13c.

It is the name of the female saint Cein the virgin, W *gwyryf*, hence *Ceinwyr(yf)* with loss of the final syllable which forms the second part of the name of the parish of Llangeinwyr. Her name stands alone in Llan-gain between Carmarthen and Llansteffan and again, with the attached *gwyryf*, as part of the name of an extinct chapel which A.W. Wade-Evans listed in his *Parochiale Wallicanum* as *Capel Cain Wyry*, a possession of the abbey of Talyllychau (Talley) which became Llangeinwyry and was later corrupted to the existing farm-name Llwyncynhwyra, whose close neighbour is, appropriately, Danycapel 'below the chapel'.

Maes-y-ward

In the 1970s a number of Bronze Age burials were discovered on the land of the farm whose name appears as Maes y Ward on modern Ordnance Survey maps. It stands near the small village of Welsh St Donats (Llanddunwyd) in the Vale of Glamorgan and was in the old lordship of Tal-y-fan.

In the published reports of these remains, some uncertainty is shown concerning the name of the property on whose land the burials were found with preference being given to the corrupt form *Maeshwyaid* or *Maesyrhwyaid* 'ducks' field', being W *maes* 'open land, place, field' + *hwyaid*, the pl. form of W *hwyad(en)* 'duck'. This is the form which appears consistently on nineteenth-century editions of Ordnancy Survey six-inch maps and is also what appears in the relevant volume of the Royal Commission's Glamorgan Inventory.

One form of the name recorded during the reign of Elizabeth I was *Mayse Heyward* and in 1579 *Mase heyward*. By 1603 it was *Maesheyarte* and in the eighteenth century the variations produced by scribes are legion, including *Musead, Mesea(r)d, Maesiad, Mosead*,

Maesyad, Marsiad and the like which continue into the nineteenth century, except for forms like *Maeseyward, Maes Ward* and *Maes y Ward* which begin to appear in parish registers, registers of electors and local government documents, until we have *Maessiward* in an 1813–14 unpublished two-inch OS map which would have been far more acceptable than the *Maes Hiad* which followed in the 1833 first edition one-inch map or the unfortunate *Maeshwyaid*.

    B.D. Harris, who made a study of the place-names of the lordship of Tal-y-fan, was inclined to see here the W *maes* + the English surname Heyward, Hayward, on the basis of the fact that John le Heyward, the lord of Merthyr Mawr in the fourteenth century, was the husband of the heiress of Richard Syward or Siward, who possessed the lordship of Tal-y-fan about the middle of the thirteenth century. This is a reasonable assumption but it is to be regretted that forms of the name earlier than 1579 are not available to prove it, because on the other hand, since this judgment can only be made on the basis of historical evidence which is not totally sufficient, is it not equally feasible to suggest that the second element in the name may have been originally the surname of Richard Siward himself in an original form such as Maes Siward, Maes Syward? The largely meaningless present day Maes-y-ward comes from an erroneous division of that form on the assumption that the two elements were separated by the W def.art. *y*.

Meiros

The farm of this name which is the topic of this note stands on the lower southern slope of Mynydd Meiros (*Mynydd Myres* on George Yates's map of Glamorgan, 1799) and a little to the east of the village of Llanharan, close to, and some distance above the residence of the eighteenth-century Powell family, Llanharan House, with its spectacular spiral stairs of cantilevered stone treads.

    Unlike names which contain the diminutive and plural ending *-os*, like the Merthyr Gurnos (pp. 88–9), Meiros has W *rhos* 'moor, heath', lenited initially, as its second element, at least two other examples of similar names occurring in the parishes of Llanegwad and Llangeler, Carms. Uncertainty arises, however, over the interpretation of the first element *mei-*. Professor Melville Richards once tentatively suggested that it could be a variant form of W *ma-* 'place, location' which occurs as a suffix in names like Gwynfa, Disgwylfa etc. or as a prefix (with its other variant form *ba-*) in names like Machynlleth, Machen, Bathafarn, Bachynbyd etc., where it probably has the extended meaning of 'plain, open country, territory'. Names which conform with this pattern are Meiarth, Meifod, Meidrum etc. and the lenited variant *-fai* as a suffix in Cil-fai, Crynfai, Gwynfai (Gwynfe), Myddfai and Pen-y-fai.

Of these various forms, Meifod is one which occurs in several places, north Wales in particular, including the lenited form after the W def.art. *(y) Feifod*, which can be seen on a signpost on the A5 road near Llangollen as Vivod (*Veyvot* 1391–3) and which seems to have retained its MW orthography, as Tomos Roberts has shown. In this form the second element is W *bod* 'homestead' and Sir Ifor Williams was of the opinion that the first element *mei-* here was based on W *meidd* 'middle, half', Lat *medius*, and seen as the second element in W *perfedd* 'middle, centre' of a place or object, so that *meifod* was construed as either a kind of 'half' or temporary abode or lodging or, as Melville Richards had suggested on the basis of Sir Ifor's definition of another of these names, Meidrum (*mei* + *drum* 'ridge, summit') as the 'centre, or central ridge or summit', *meifod* could be a lodging or dwelling place in a position, as it were, half way between the summer dwelling, the *hafod*, and the old residence, the *hendre*. It is an aspect of this sense which is favoured here, but it is difficult to be specific.

Was Meiros a portion of moorland or heath in a central position and, if so, in relation to what? Does it convey the meaning of the 'middle or centre of the *rhos*' or does it refer to the 'open' land which constituted the centre of the *rhos*? The collected forms of the name include *Mayros* 1611, *Miras* 17c., *Meiros* 1701–2, *Maeres* 1833, *Meyroes farm c.* 1840, *Mairos* 1884, but its composite form suggests a much earlier period of origin.

In 1611 it was the residence of a reasonably prosperous man, William Thomas, who received £600 from Sir John Stradling in a land transaction and it may be pertinent to ask whether Meiros was the predecessor of Llanharan House as the 'great house' of the parish of Llanharan. The Powell residence was an eighteenth-century structure with later additions. Its site was known as Ton-y-gof 'the blacksmith's plot' and was not acquired for the purpose until *c*. 1694. The earliest reference to the house in surviving documentation is no earlier than 1750, as far as is known at present.

Y Mera

The town of Neath was built on a convenient site which afforded a crossing of the river Neath and it was a wet and marshy site on the upper reaches of the estuary before it widens down to the sea. This is confirmed by a number of minor place-names which are recorded in land surveys and the like, some of which are now lost, like the Western Moor, Castle Marsh, or the Latts (a parcel of wet land which could be a ME plural form of the modern *lath* in the sense of 'a long strip of land').

Towards the far end of Water Street (another such name), where the present Victoria Gardens are situated, there was a portion of unproductive land where, by the seventeenth century, a number of workmen's cottages had been raised, which included the Miners' Row. This was the Mera.

Gwŷr y Mera 'the men of the Mera' were a diverse mixture of humanity who had come to work for Sir Humphrey Mackworth and other employers in the area, a distinct social group within the population of the town. So were their wives, we are told. Merched y Mera were the Amazons of Glamorgan, according to D.R. Phillips.

Earlier, quite a strong-running stream flowed through the Mera to the Pwll Cam 'the crooked, or bent pool', and the indications were that this pool was all that remained of a much larger spread of shallow water which occasionally flooded the area. As the town was an English foundation, in essence, it is not inconceiveable that this expanse of water should have been called by the ME term *mere* 'a pool, a marsh', and it is essential to appreciate that the final *-e* would have been enunciated as a final 'dark' syllable. This is very probably what accounts for the Welsh-speaking elements of the population retaining the syllable in the form *mera*.

A somewhat similar development can be seen in a reference to Llan-gors lake, Brecs. In 1725 it is referred to as the great pool called *Mara Llangors* and it is clear that the standard reference to the church of Llan-gors in the episcopal charters of the diocese of St David's from the twelfth century onwards is as *ecclesia de Mara* which becomes *ecclesia de Langors* by 1234, and the lake itself as *la Mare* in 1327.

It is likely that *mara* in these excerpts is a Latinised form of the English *mere* with a vocalic variation to be compared with *mera*. Indeed, Llan-gors lake is later referred to as 'a body of water ... well known by the name of Welsh Pool or Brecon Meer' (1793).

There was a *Panwaun y merra* in Rhyndwyglydach in 1847 (W *panwaun* 'flax moor').

Merthyr Mawr

The value of comparative work in the elucidation of Welsh place-names has not always been fully appreciated. The name of this secluded village in the Vale of Glamorgan is a case in point as it can be effectively interpreted by comparing it with that of Llanofer in Gwent (more familiar, unfortunately, in the form Llanover) despite their apparent incompatibility.

As in the majority of names beginning with W *llan* (but certainly not all) the prefixed element is followed by the name of the 'saint' to whom the original cell was consecrated and many earlier documented forms of the name Llanover seem to favour *Llanfofor* (*Lanmouor* 1245–53) as the archetype, leading to *Llanofor* in common parlance by loss of medial -*f*-. Nineteenth-century antiquaries who were aware of the process of lenition of the initial consonant of the element following the n.f. *llan* in a name assumed, therefore, that the spoken form -*ofor* was the lenited form of a personal name which began with *g*-, lenition of which caused its disappearance as in *llan* + *G(w)rwst* > Llanrwst, *llan* + *Gwrin* > Llanwrin etc. Hence the appearance of a 'saint' called *Gofor* whose name appeared to feature in Llanofor.

Consequently a well in the grounds of the mansion Llanofer Fawr, which was presumed to have a connection with the local parish church and was called Ffynnon Ofer, had its name transferred to Kensington Gardens in London by Benjamin Hall, Lord Llanofer, and anglicised as St Govor's Well.

However that may be, attention should have been given to earlier forms of the name Llanofer, especially those which appear in ecclesiastical documents of the thirteenth and fourteenth centuries. These indicate firmly that the 'saint's' name which occurs as the second element was not Gofor but Myfor, so that the original form (in modern orthography) was *Llanfyfor* which became *Llanfofor* in spoken forms by vowel assimilation and then *Llanofor* (*Lannouor* mid 13c, *Llannovore* 1353) with the variant colloquial form Llanofer (*Llanover* c. 1291) which has survived.

In OW orthography the personal name Myfor would have appeared as *Mimor*, and in the *Liber Landavensis* there occurs the name of an early Christian settlement in the Vale of Glamorgan in the form *Merthir Mimor, Merthir Myuor, Merthir Mouor*, that is, Merthyr Myfor, which contains the same personal name as Llanfyfor.

It is the subsequent influence of common parlance by a mixed linguistic population which has caused the indistinct pronunciation of the element *-myfor* in the form Merthyr Myfor, again helped by the loss of the medial *-f-* > *my-or*, thus giving rise to a form which is approximately the way the W adj. *mawr* is pronounced by speakers of English and represented in documentation, both as an adj. and as a surname, by *maur(e), mawr(e)* and possibly *mawer*. This occurred quite early, as is attested by forms such as *Marthelemaur* 1146, *Merthur Maur* 1262, *Marghilmaur c.* 1291 etc., but it was certainly taken to be the W adj. 'big, great' in Welsh sources before the end of the fifteenth century. This led to erroneous interpretations (some quite recent) of the element *merthyr* in the name as if it meant 'martyr'.

If we accept that in modern times the exact difference in the meaning and significance of the Welsh elements *llan* and *merthyr* has been somewhat eroded, it could be said that Merthyr Mawr and Llanofer are not far from being synonymous.

Morlanga

A farm which bears this name stands on land which slopes gently down to the banks of the river Ely (Elái) in the parish of Peterston super Ely (Llanbedr-y-fro, or Lanbedr-ar-fro, as the late Professor G.J. Williams would have reminded us). Although not exactly on the river bank, a fair portion of the land would have been subject to the occasional overflow of its waters, being subsequently wet and marshy. This affords a clue to the meaning of the name although it seems to have caused problems to several investigators in the past, one attempt having been made to ascribe to it a Scandinavian provenance.

It is not particularly well documented, the earliest forms found being *Morlange* 1613–14, *y Morlangey* 1654–5, *morellange* 1687, to be followed by forms which appear to interpret the first element as E *moor*: *Moorelangey* 1712, *Moorlonge* 1738, *Moor Langey* 1790, *moor langha* 1824, later to be restored as *Mor(e)langa* 1846 etc., the final *–a* being wrongly interpreted as the shortened colloquial form of the pl. ending *-au* and 'restored' on the 1833 Ordnance Survey one-inch map as *Morlangau*.

There seems to be no reason to doubt that the first element is the W n.f. *morlan* < W *môr* 'sea' and *glan* 'edge, bank' (as in its equivalent *glan y môr* 'sea shore') referring especially to wet, boggy land in such a location but having become an established term for low-lying open land, inland, which is inclined to be wet and marshy, especially near rivers, in much the same way as the well-established W *morfa* '(salt)-marsh, fen, moor', originally < *môr* + the commonly appended *-fa* < *ma* 'place', but in common use in inland situations. The attested forms which have E *moor* substituted for the first syllable *mor-* therefore show a measure of appreciation of the meaning.

The appended *-ga* in *morlanga* can be understood as an initially lenited colloquial form of W *cae* 'field' > *cā*, but with the vowel shortened in an unaccented final syllable. Corroboration comes in the form of the name of another farm in the same parish, Gwernga, which is W *gwern* 'swamp, marsh' + *cae* (Gwern-y-gae on the map).

There was a farm in Merthyr Tudful before industrialisation called Morlanga (*Morelangley* 1630, *Morlanna Farm* 1800) and David Watkin Jones (*Dafydd Morganwg*) in 1874 testified to the existence of a cot on its land called Y Forlan. Further, in the parish of Ystradowen in the Vale of Glamorgan there was another farm of the same name which is recorded as *Burlonga* in 1760 and *Bodlonga* on the 1833 Ordnance Survey one-inch map, but is *Morlonga* early 19c., *Moorlangha Farm* 1824, now Ffald Farm, and is situated not far from Morfa Ystradowen.

Nant y Cesair

Some rivers and streams are named because their waters are cold or warm. R.J. Thomas, our greatest authority on Welsh river and stream names, has noted numerous examples throughout the country. Nant Twymyn, Mont., is tepid (W *twym* 'warm, tepid'), Y Nant Boeth which flows into the Afan has W *poeth* 'hot, warm', and in several parishes in Glamorgan there is a Ffynnon Dwym, so-called because of the lukewarm water contained in the wells so named.

On the other hand, Oernant, Nant Oer and Oerddwr (with W *oer* 'cold') are common, and Odnant runs down Newydd Fynyddog, Mont. (with W *ôd* 'snow') which corresponds to Nant yr eira (with W *eira* 'snow') in several places. It is probably W *ia* 'ice' which is the base of Nant Iaen in the Rhondda Valley. Near Bethesda, Caerns., there is Afon Genllysg (with W *cenllysg* 'hail') and near Creigiau, in the parish of Pen-tyrch, there is a Nant y Cesair. W *cenllysg* and *cesair* are synonymous.

However, a change befell Nant y Cesair in one respect, for at the side of the road from Creigiau to Efail Isaf there stands a well-known public house and restaurant, the Caesar's Arms. The stream runs down from higher ground to the east, under the road and on past the gable end of the hostelry.

In this district the Welsh colloquial form of *cesair* is *cesar*, and the stream's name is recorded as *nant y kessar* 1570, *Nant y Cessar* 1796 and *Nantycessar* in a remarkable Welsh-language valuation of the parish of Pen-tyrch in 1824. However, it also happens that *Cesar* is the usual Welsh form of the Latin *Caesar*. For instance, the rendering of the phrase 'render unto Caesar that which is Caesar's' in the Gospels is *rhoddwch yr eiddo Cesar i Cesar* in the Welsh Bible, and an oft-quoted phrase in daily converse.

It has been said that the stream was once referred to as Caesar's

River, but whether that is true or not there can be little room to doubt that such a coincidence of the oral transmission of the name of the stream, *Cesar*, and the Welsh version of Caesar, together with the close proximity of the stream to the public house, inspired the naming of the hostelry as Caesar's Arms. Thus was presented a golden opportunity to name the hostelry after an figure of note and a high-ranking Roman figure in all his glory is consequently depicted on the sign outside.

Let this be a warning to those who would search indiscriminately for Roman remains in the parish of Pen-tyrch!

Nantyffyllon

Some place-names indicate the value of the land which is named by including in their forms the sum of money once paid as rent for the holding. In the parish of Aberdâr (Aberdare) Tir y pumpunt is recorded from 1570 (W *pum(p)* 'five' + *punt* 'pound'); in Swansea, Tir Deunaw (*Tyre Doynawe, Tir y doynaw* 1650, W *deunaw* 'eighteen' from W *dau* 'two' + *naw* 'nine' for 18d. or 1s. 6d.), and one wonders what value may be implied in Tir-y-pwrs (*Tir y purse* 1666, W *pwrs* 'purse'), also in Aberdâr.

A further interesting example where the reference may be directly to a coin rather than a sum of money is Nantyffyllon near Maesteg, however unlikely this would seem at first glance. Originally it was the name of a stream (W *nant*), a tributary of the Llynfi river, and the earliest attested forms of the name are clear in their meaning: *Nant firlling* 1570, *nant ffyrllinge* 1630, (*Blaen*) *Ffyrlling* 1633, *Nant Ffirlling, Nant Ffyrlhing* 1695–1709, *Nant Fyrling* 1736, *nant Furling* 1740.

The form *ffyrling, ffyrlin*, is a colloquial form of W *ffyrlling*, borrowed from ME *ferling* 'farthing' although it may be doubted whether a direct connection with the coin is implied. What is implied is probably an applied and figurative sense of something that was small or insignificant, as the farthing was the coin which was lowest of all in value. This must indicate the stream's status in the estimation of the population of Tir Iarll.

Its name survived, however, which is more than can be said for some of the Llynfi's other tributaries, possibly because of its use as an element in the name of the uncultivated and wooded region which took its name, *Fforest Nant Ffirlloige* (sic) 1588 etc., and by 1787 it appears that a holding in the area had been enclosed, *Tir Nant Ffirlling*, which had become *Nant y fferling Farm* by 1846.

Since the evidence of the early forms are reasonably conclusive in this instance, it can be assumed that the present form Nantyffyllon evolved from a colloquial form of *Nant (y) ffyrlling* by loss of *-r-* in the consonantal group *-rll-* and the vowel of the last unaccented syllable becoming indistinct.

Yr Olchfa

This is a Welsh term which occurs as an element in place-names and as a simplex form in the north and south of the country, two examples occurring in the neighbourhood of Swansea.

Near Sketty (Sgeti) in 1624, *y tyr wrth ryd erolchva* is recorded (= *y tir wrth ryd yr olchfa* 'the land by the ford of *yr olchfa*') and the stream which is fordable at that point is a brook called the *Olchva*, 1744. Further, in the upper region of the Lliw valley stand the two farms of Blaenyrolchfa Fawr and Blaenyrolchfa Fach but although the names, as farm names, are on record in the seventeenth century and thereafter (*Blaen yr olchfa* 1687, *Blaen yr Olchva* 1739, 1764 etc.) Rice Merrick in 1578 notes clearly that *Blaen yr Olchva* is the name of a hill, the location of the source of the river which he calls *Llugh ye lesser*, now the Afon Llan (for which see p. 155.), 'Llugh ye lesser ... springeth out of a mountain called Blaen yr olchva in Llangyfelach ... ', this being the high ground at National Grid reference SN 677052 which is called *Mynydd llaen olchva* (sic) in a survey of 1650, a *Pant-yr-olchfa* being shown in the vicinity also on the 1884 six-inch Ordnance Survey map.

In both these examples *yr Olchfa* occurs as a river or stream name, but it would appear that the streams acquired their names because of the existence somewhere along their courses of a location which was termed Yr Olchfa. The form is essentially self-explanatory, being the lenited form of *golch-* after the W def.art., the stem of the W verb *golchi* 'to wash' + the common ending *-fa*, from W *ma* 'place, location', in the sense 'washing place, washery'. GPC also adds *baddon* 'bath', though any modern connotation of that term can be discounted in this connection.

In fact, the Welsh term *golchfa* reflects an aspect of the agricultural scene and refers specifically to the traditional use made of streams and rivers to wash or dip sheep before the advent of modern constructions for the purpose. In his discussion of the names Yr Olchfa, Llyn yr Olchfa and Golchfa 'Ralltlwyd in the valley of Abergeirw, Merionethshire, the late D. Machreth Ellis succinctly described what took place (in translation):

> ... the sheep was thrown into a pool which had been dammed in a river or stream ... and then, as it turned to straighten itself in the water and swim out towards the shallow end of the *golchfa*, the weight of the water would loosen the fleece.

He goes on to comment on the fact that a number of farms in a neighbourhood would take advantage of the facility and would join together to make the washery at a convenient spot by constructing a weir across a stream to form a pool.

As far as is known, the term *golchfa* has not been attested among the names of the parish of St Andrews in the Vale of Glamorgan but it is possible that it is the construction of such a washery which is implied in the name of the farm Argae 'weir, dam, sluice' on a stream near the parish church.

Pantaquesta

On first sight this name has the appearance of a name which looks unlikely to have a Welsh origin. In this form on current Ordnance Survey maps, with its -*qu*- combination, it is the name of a farm and, of late, a small group of houses close to the road which runs under the M4 motorway from the village of Meisgyn (Miskin) to Hensol Castle, but a perusal of its earlier recorded forms show clearly that it is indeed a Welsh name. The earliest forms are: *Pant Ygwestay* 1558–9, *Pant y gwestaye* 1574, *Pant y Gwestay* 1627, *Pant y Gwestey* 1641, *Pant y Gwesty* 1660–1, *Pant y gwesta* 1768, which point to an original Welsh form (in modern orthography) Pantygwestai.

By 1838 provection of the medial -*g*- to -*c*- had occurred and the reduction of the final -*ai* to -*a* in common speech to give the form *Pantycwesta*. This in turn was 'restored' to *Pant y cwestau* on the first edition one-inch OS map of 1833 in the mistaken belief that the final diphthong, which had been reduced to its dominant vowel -*a*, was originally the common W pl. ending -*au*. In addition, the Welsh combination of *cw*- was presented by non-Welsh scribes as -*qu*-, as in the present map form. This had already occurred in a will of 1793, *Pantyquesta*, the later medial -*a*- appearing for the unaccented W def.art. *y* in the form Pantaquesta.

As is well known, many names of settlements originate as descriptions of their location and are topographical in character, in this case that of the farm. When homesteads are situated near prominent features, for example, the name of that feature is adopted as that of the settlement with, perhaps, an additional element to show possession, nature, extent etc. However, it may well be that in the present case the location was already named *Pantygwestai* before the middle of the sixteenth century, were there earlier recorded forms to prove this. Two considerations incline one to this view.

The first is the fact that the document of 1574 in the Talbot of Hensol family estate papers quoted above refers to an enclosure named *Pant y gwestaye*, 'a close' and not a homestead as such. The use of W *pant* 'hollow, depression, valley' as the first element in the name is appropriate to the location, with a large pool at its lowest point.

Secondly, the second element is the W n.m. *gwestai*, preceded by the W def.art. *y*, which now has the meaning 'guest, visitor' but which does not accord particularly well in a topographical context in that sense as a qualifier for *pant*. Far more convincing is *gwestai* in the more specific and specialised meaning which it had in the old Welsh Laws as the name of a court official in the royal household, 'a provisioner'. In the old legal texts 24 officers are listed and named together with 12 *gwestai*, these being the officials responsible for collecting the *gwest* or provisions for the maintenance of the king (or lord) and his court. Such men were probably prominent in their own localities: one commentator goes so far as to suggest that they were 'dignitaries' of the localities which the court periodically visited. The 'provisioner's hollow' may in this case preserve the memory of land granted to such an official, possibly one of the retinue of a Welsh lord of the commote of Meisgyn in pre-Norman times.

Pantygwydir

It is unfortunate that the street names Gwydr Crescent, Gwydr Square, Gwydr Mews and Pant-y-gwydr Road in the Uplands area of Swansea appear in this form on maps and occasionally in printed works. Presumably they originate in an attempt to make some sense of the element which appears as *gwydr* in these versions, for *gwydr* is the Welsh word for 'glass', although it would hardly seem to have had any relevance in this context. The correct form is *gwydir* and the street name which contains the original form on which the others are based is Pantygwydir Road.

Until the first half of the nineteenth century, as is well known, the present urban area in which these modern names appear was well outside the limits of the old borough of Swansea. It was a rural area of woods and fields extending down the slopes from the high ground of Townhill and dotted with dwellings and farmsteads, one of these being Pantygwydir, situated in the vicinity of the present Pantygwydir

Road. The tithe map of 1843 shows it to be within the width of a field or two to the north of The Rhyddings (pp. 164–5) and to the west of Bryn-y-môr (now the Stella Maris Centre).

The name is descriptive of its location at the time of its original application, a 'depression, hollow', W *pant*, and is attested in a fair selection of recorded forms which include *Pantygwyder* 1599, 1626, 1651, *Pantguydir* 1650, *Pant Gwider* 1736, *Pantywidir* 1776, *Pant Gwydir* 1798, *Pantgwider* 1843, *Pant-y-Gwydwr* 1852 where uncertainty about the second element becomes more apparent, and the implausible *Penquidar* of George Yates's map of Glamorgan of 1799.

As noted above, the second element here is *gwydir* which, like *gwedir*, is a variant form of the W n.m. *godir*, which has the modifying prefix *gwo-*, *gwa-* + *tir* 'land' to give the meaning 'low ground, depression, slope' (in contrast with *gorthir* 'high ground' which has the intensifying prefix *gor-* + *tir*). It is cognate with Gaelic *fothair* 'slope' in names like Fetters, Fife, or Foithear (Foyers), Inverness, etc. GPC notes that the parallel form Gwydir, Gwedir, the name of the home of the celebrated family of the Wynns of Gwydir in the Conway Valley has the specific sense of 'low-lying or lower ground', and as the second element of Rhosygwidir, Pembs., it probably occurs in the same sense. The combination of this element and *pant* has the appearance, therefore, of a tautologous form, but the reference may well be to a hollow or depression in what is, after all, the 'lower' ground of that which slopes from a higher elevation and which cannot be identified in modern times.

This modest homestead came into the possession of the Tennant family and was expanded *c.* 1860 for John Crowe Richardson of Glanbrydan, Maenordeilo, Carms., but was demolished in 1908.

Pebyll

A friend drew attention to the name Pebyll, today the plural form of W *pabell* 'tent, temporary dwelling, shelter' etc. on the ridge of Mynydd Blaengwynfi in the northernmost area of the Ogwr valley and wondered how such a connotation could have been applied to a place in such an exposed and windswept spot.

This is to be explained partly by the fact that although *pebyll* is a modern plural form it was not always so, for in fact it is the regularly derived singular form from the Latin *papilio* and plural forms are attested which were formed in the normal manner by adding plural endings, such as *pebyllau, pebyllion* and *pebylloedd*. In this case, however, there also occurred the formation of a new singular form *pabell* by analogy with other W nouns which have the vowel sequence *-e-y* in their plural forms but *-a-e* in the singular, such as *cragen* 'shell', pl. *cregyn, astell* 'plank, lath', pl. *estyll*, etc. *Pebyll* was likewise popularly regarded as a plural form on this pattern with an assumed singular form *pabell*, which acquired common currency in both the literary and the spoken language. Further plural forms were also created from the 'new' singular form *pabell* by the addition of the normal plural endings, such as *pabellau, pabelli* and *pabellion*.

The location referred to here on the exposed ridge of Blaengwynfi Mountain is a Bronze Age ring cairn with its stones upcast from a central depression. In popular estimation these remains no doubt suggested some kind of edifice, shelter or encampment of an earlier age and were given the name, preceded by the W def.art. (now lost) in all probability, *Y Pebyll* (with which the use of W *gwersyll* 'encampment' for similar locations may be compared). It occurs in this form near Llangrallo (Coychurch) as the name of a small cavern Ogof y Pebyll (W *ogof* 'cave'), although Pebyll-y-brig in the parish of Llantrisant, 'the shelter on the summit', was a small cot.

Normally regarded as a masculine noun, evidence can be found of *pebyll* as a femine noun, as is indicated by the lenition of leading consonants of following elements in names like Penlle'rbebyll on Mynydd Pysgodlyn, near Pontarddulais (another Bronze Age ring-work) and, notably, Cilybebyll where the significance of *pebyll* is obscure (as obscure as the true meaning here of the first element, *cil*). It may well be that some semantic development accounted later for its use to denote simply a residence of a temporary nature like a hovel or cot.

One scholar derived the name Peebles in Scotland, through an intermediate English form, from the W *pebyll* rather than from a Gaelic root, and suggested that the term can be likened in its meaning to E *shiel* (*shieling*), ME *schele* 'hovel, hut, cot'.

Pencisely

This is the modern form of a name which appears on maps and also in the street names Pencisely Road, Pencisely Avenue, Pencisely

Crescent and Pencisely Rise between Llandaf and Victoria Park near Canton, Cardiff, the second element after *pen-* being pronounced as if to rhyme with E *nicely*.

It appears that the visual impact of the terminal *-ely* in written or printed forms of these names has caused some to ponder the proximity of the river Ely (afon Eláí) here, for Pencisely Road, the more important of these streets, runs westwards down Ely Rise, close to Ely Bridge (Pont Treláí). Indeed, one ingenious explanation of the name in current circulation is that it is a telescoped form of the combination of Welsh elements *pen-cae-is-(e)láí*, 'the end of the field below the Ely', but no recorded forms of the name can be produced to support this contention.

It is probable that W *pen* here means 'high ground, a height' rather than 'the end of', this being acceptable when the configuration of the land on which the homestead which first bore the name is taken into consideration. This was elevated open ground and the name may well go back to a period before the middle of the sixteenth century because the property which bore the name had already been divided by 1543 when *Pencysle Yssa* (that is, Pencisely Isaf 'lower Pencisely') is recorded without reference to a probable Pencisely Uchaf 'higher Pencisely'. It may well be that the former became Pencisely House at a later date and the other Pencisely Farm, both of which were demolished by the 1930s.

As other names in the locality tend to confirm, the location is a ridge of land which rises from the west, the Ely Rise noted above being one example, and extends to the east, along which runs the present Pencisely Road descending gradually in the direction of Canton (Tregana) in the vicinity of Penhill—a self-explanatory hybrid name of synonymous Welsh and English elements, but obviously referring to rising ground. The properties which became Pencisely House (*Pensishly House* 1869, *Pen-sisli House* on the six-inch Ordnance Survey map of 1886) stood about halfway between the two extremities and opposite the present Pencisely Rise (yet another indication of higher ground) which leads down to Victoria Park. It has been claimed that the gate-posts of 155 Pencisely Road at the present time were those of the old Pencisely House.

Between the 1543 and 1886 forms noted there also occur *Pencissly* 1612, *Pensisley* 1722, *pen S(h)ishley* 1763, *Penshisly* 1762, *Pencicely* 1779 and *Pensishly* 1833, all of which strongly suggest that the second element in the name is the female pers.n. Cicely, the variant form of Cecily and Cecilia in one of its common popular forms, Sisley. It is also interesting to note a feature normally seen in spoken Welsh in south Wales displayed in some of the recorded forms, namely the change by which *s* becomes the palato-alveolar *sh* after *i*, as in *disgwyl > dishgwl, meistr > mishtir, pris > prish* etc. to produce the forms *Pensishly* 1763, 1833 and the metathesised form *Penshisly* 1762. Examples of the comparative W adj. *isaf* 'lower' in the form *isha(f)* are numerous in this volume. It is also relevant to note that Eastbrook in the parish of St Andrews appears as *Ishbrook* in 1884 and, similarly, an original Easton as *Ishton* 1813, now Yniston in Leckwith (pp. 221–2).

The personal name, therefore, is an indication of association or ownership, but the identity of the original Cecily is not certainly known.

Pen-cyrn

The farmhouse of this name stands near the old road from Ystradowen to Aberthin and Cowbridge in the Vale of Glamorgan, near Trerhingyll, and although the earliest form of the name seen hitherto is *Penkyrne* 1603, the Royal Commission on Ancient Monuments in Wales cannot date any structural features of the building earlier than the end of the seventeenth century. Included among the remaining recorded forms are *Penkirn* 1690, 1786, 1799, *Pencirn* 1718, *Penkyrn* 1790, with *Pencurn Wood* 1788 which appears in the Welsh form *Coed Pen Cyrn* on the six-inch Ordnance Survey map of 1884.

The present form, Pen-cyrn, suggests that the second element is W *cyrn*, the plural form of the W n.m. *corn* 'horn', or possibly the old Welsh dual plural form, used in an adapted sense to refer to a mound or cairn, being descriptive of the location of the original farmstead so named. However, if this is the case, the use of a plural form here is unclear since there seems not to be a single feature of that nature in the vicinity of the slope on which the building stands, namely a bare ridge running southwards past the gable end. This also suggests that the first element, W *pen*, cannot be interpreted as meaning 'the end of' but possibly as merely referring to the higher ground of the ridge.

There is another much discussed W n.f. *cyrn* or *curn* (the form varies) which is a singular noun and which means 'a heap or mound' often in the form of a pyramid or cone. It occurs in the names of mountains in particular, Y Gurn Goch, Y Gurn Ddu, Y Gurn Las in Gwynedd (see p. 88). Since these natural features are isolated heights and do not exist in 'pairs', Sir Ifor Williams rejected the idea that *curn, cyrn* may be the old dual form of *corn*. Further, the singular *cyrn* seems to have its own plural form which exists in the mountains of Snowdonia as Y Cyrniau. Even so, there may be some substance in the idea put forward by Egerton Phillimore that *cyrn* could have been 'a specialized use of the plural of *corn* in a singular sense' noting the fact that a diminutive form *curnen, cyrnen* also for 'a heap, mound, stack' was formed from it. He notes further that *curn, cyrn* does not necessarily have to refer to lofty conical or pyramidal peaks, that 'it is found elsewhere, used apparently of mere rocky tumps or tops of small hills' noting Pen-y-gyrn, Mont., and Gyrn near Rhuthun, Denb., to which can be added Mynydd Pen-cyrn to the west of Llanelli, and Y Gyrn near the Storey Arms, both Brecs. The moderate elevation of Pen-cyrn at Ystradowen may have been sufficient to merit this modified application of the term to its position.

The remaining possibility, of course, is that the feature which gave the property its name has now disappeared, especially if it was originally man-made. At no great distance is the unfinished motte to the west of the church at Ystradowen which may have been intended to be either the centre of the old lordship of Tal-y-fan by the family of St Quintin in the twelfth century, or that of the Norman lordship of Cherleton (1148–83), but this is an unlikely candidate for consideration.

Pendyrys

Since there is no place in the Rhondda Fach valley which bears this name, it was not unreasonable for a friend who is a member of the renowned Pendyrys Male Voice Choir to ask why it should be so-named.

The answer lies on the high ground of Cefn Gwyngil in the upper region of the valley and on the eastern side of the river. This is land which is in the parish of Llanwynno and over which the road from Blaenllechau climbs purposefully towards the site of the old church of that parish. This was the broad location of a portion known as Pendyrys, a compound of W *pen* 'head, end of', but which can also be found in reference to a ridge or projection, sometimes rocky or craggy, and the W adj. *dyrys* 'wild, rough, uncultivated: tangled (with undergrowth), thorny, matted'. This is a form (with the addition of an epenthetic vowel) based on the stem *drys*, as seen in the verb *drysu* 'to entangle, encumber', the noun *drysni* 'intricacy, perplexity, confusion', and the first element of the well-known district and castle of Dryslwyn 'tangled bush, thicket, bramble-brake' in the parish of Llangathen, Carms. or its fairly common counterpart with the elements reversed, Llwyndyrys.

By 1778 the holding had been divided into two portions, Pendyrys Isaf (W *isaf* 'lower'), vestiges of the walls of the homestead being still visible by the side of the road to Llanwynno, not far from the present farmhouse of Blaenllechau, and Pendyrys Uchaf (W *uchaf* 'upper') which, as far as is known, no longer exists.

The parcel of Pendyrys land, however (either the original complete holding or the 'lower' of the two moieties), extended down the valley to include the slopes which descend behind the present area of Tylorstown, and what occurred during the coal-mining boom of the nineteenth century was that Alfred Tylor of Newgate, London, bought the rights to the minerals which lay under the Pendyrys land in 1872. He mined successfully and was producing nearly a quarter of a million tons of coal in the year 1893 when the mine was sold to new proprietors.

This was the colliery which was named the Pendyrys colliery, as might have been expected. What is also apparent is that the location of the community created by the colliery was named after Alfred Tylor, Tylorstown, but that the community's male-voice choir was named after the place of work of most of its members.

The remembrancer of the parish of Llanwynno, Glanffrwd, maintained that Pont-y-gwaith was the old name of Tylorstown, but some doubt may attach to this claim. Further up the valley, it was on the land belonging to the homestead of Blaenllechau that the shaft of the local colliery was sunk, but the name was retained as the name of the colliery and the village which was inhabited by the community thus created although the community of the nearby Glyn Rhedynog adopted as its name the literal English translation, Ferndale.

Penhefyd

During periods of anglicisation of place-names in the Vale of Glamorgan there came into existence, inevitably, a type of name which was a hybrid form of synonymous elements in Welsh and English, or vice versa. Examples of this kind are Brynhill, Garnhill, Maeslon (W *maes* + a form of E *land*) or Bryndown, near Dinas Powys, W *bryn* + E *down* in the simple sense of 'hill' in ME, which was erroneously 'restored' to its present form Bryn-y-don. There is a case for regarding Penhefyd in St Fagans as a name formed on the same pattern. It is that of the farm of the manor on the edge of Pentrebach which stands on a little knoll of no great height but sufficient to make it noticeable in its surroundings. Nearby stands the pavilion of St Fagans cricket club.

It can hardly be claimed that the second element in the name is the Welsh adverb and conjunction *hefyd* 'also, besides' etc. for that form does not appear among the earlier forms recorded until the beginning of the nineteenth century as *Penhyfed* on the first edition one-inch Ordnance Survey map of 1833 which was changed to *Penhefyd* on the 1885 six-inch map. The earliest forms are as follows: *Penhevette* 1585–6, *Penhevett* 1587, 1608, 1675, *Penhevet* 1609–10, *Penheved* 1766 and *Penhevad's farm* 1765, all of which suggest strongly that the second element, with W *pen*, is ME *hevet*, *heved*, OE *hēafod*, which is ModE *head*.

Although OE *hēafod* has a number of adapted topographical meanings in place-names in England it is accepted that the most common is 'rising ground, hillock, hill', no matter how modest its proportions, provided it stands out fairly prominently in its surroundings. Margaret Gelling says that it is 'sometimes used of very low projections', citing the name of a place in Berkshire, Hatford (where the first element is originally *hēafod*) on a very slight elevation above the marshy course of a river. W *pen*, the first element of Penhefyd, can also be regarded in the same light, to the extent that what had become *Penhevet* in ME was clearly a hybrid form of virtually synonymous elements in Welsh and English to refer to the original homestead's location. Although meaningless in this context, the representation of the second element as W *hefyd* is an obvious analogical development where phonology rather than semantics is the governing factor.

Support is available for this interpretation from the existence of a similar hybrid example within the space of a field or two from Penhefyd. About a field's width on the eastern side of the road which passes through St Fagan's (Crofftygenau Road, p. 55) past Penhefyd, there stood a small homestead which no longer exists but which, according to the Plymouth estate papers, was once part of the same property, namely *Pendown* 1766, 1790, *Pen-y-down* 1802. Here again is a name which has W *pen* + E *down* (OE *dūn*) in the same sense as that noted in Bryndown above.

Penlle'rbrain

Between Gendros and Fforest-fach on the northern limits of Swansea there is a hill which today bears the name Raven Hill and the evidence of earlier recorded forms seems firmly to support the contention that this is a free translation of the Welsh form Penlle'rbrain. No English form of the name appears to be available before the middle of the nineteenth century by which time *Ravenhill House* and *Ravenhill Park* are recorded, and two street names in the vicinity today perpetuate the memory of crows, not ravens, namely Rhodfa'r Brain 'crows' terrace' and Ffordd y brain 'crows' street', W *brain* being the plural form of *brân* 'crow' rather than 'raven', which is *cigfran* in Welsh.

The earliest recorded form of the name of the hill is *Tyle'r brayn* 1650 for what should more correctly be *Tyle'r brain* (the spelling of the second element, *brayn*, being an indication of a non-Welsh scribe's reading of W *brain* to rhyme with E *brain*) with W *tyle* as first element. This is the common south Wales word (west of the river Tywi to the Wye) for 'a hill, hillock, rising ground, slope' which becomes *tyla* and *tila* in the spoken language and is probably cognate with Ir *tulach*, of similar meaning and which appears as *tully-* or *tulla-* in numerous place-names in Ireland: Tully, Tullyrone, Tullaherin, Tullagh etc.

The hill *Tyle'r Brain* was so named, therefore, probably because it was made conspicuous by the crows that frequented it. The addition of W *pen* as a prefixed element here may not specifically denote rising ground, which would make it a tautologous addition to *tyle*, but is used in the alternative sense of either 'the end of' or 'the top, summit of', the recorded forms being *Pentyle y brain* 1600, *Pentylarbraene* 1611, *Pentylau'r Brain* 1720, *Pentyllebrain* 1759, *Pentila(r)-*

brain 1764. It is a syncopated form of this which later produced the present Penlle'rbrain (*Penller Brain* 1799, *Penlle'r-brain* 1830) aided, no doubt, by the existence of the common W *lle* 'place' which was easily substituted for the second syllable of the original *tyle*, the normally accented first syllable of *tyle* having become unaccented by the addition of the prefixed *pen*.

However, another variant form of the name seems to make a fleeting appearance in the eighteenth and early nineteenth centuries, namely *Pen Llwyn Braen* 1729, 1760, *Penllwynbrain* 1754, 1838, 1854. It is true that there seems to be no means of ascertaining whether or not *Penllwynbrain* and Penlle'rbrain refer to exactly the same location, but it may be significant to note that the general area in which both occur is that which is referred to in documentation of the early fourteenth century as *Crow Wood*: *Crowewode* 1305, *Crowode* 1319. The *-llwynbrain* of Penllwynbrain, with W *llwyn* 'bush, grove, copse' for 'wood', is the more accurate rendering of this early name, the crows, W *brain*, being the common element.

A further point to be noted concerning the form Penlle'rbrain is that it is not the only name of this type in Glamorgan, with the medial *-lle'r-* preceded by *pen*, which may well have influenced the reduction of the original Pentyle'rbrain to its present form. The others are Penlle'rgaer, Penlle'rcastell, Penlle'rbebyll (for *pebyll* see pp. 138–9), Penlle'rfedwen (*bedwen*, diminutive form of *bedw* 'birch'), and Penlle'rneuadd (*neuadd* 'hall'), but it would be wrong to explain these names as names, like Penlle'rbrain, which have been reduced from original forms with W *tyle* as a second element. Hitherto, an inspection of their earlier recorded forms has revealed no trace of *tyle*.

Penllwyn-sarff

Like Bedwellty and Rasau, Penllwyn-sarff is in Monmouthshire but it is sufficiently close to the border with Glamorgan to allow its inclusion in this volume, if only because of the considerable number of enquiries which have been received as to its meaning.

Just over the Rhymni river on the left of the road from Ystrad Mynach to Blackwood (Y Coed Duon) in the Sirhywi valley, and in the vicinity of Pontllanfraith, the ground rises perceptibly and is the site of an urbanised area which bears the name of Penllwyn. Its centre is the hostelry of the Penllwyn Arms standing prominently on the ridge, a many-gabled structure of handsome appearance. It is the restored house of Penllwyn-sarff in the parish of Mynyddislwyn, the home of a branch of the well-dispersed Gwent family of Morgan in the sixteenth century, the house having been said to be 'once a fair mansion, but now almost in ruins' in 1906. It appears as *Penllwyn-sarph* in some documents, having the consonantal digraph *-ph*, the phonetic equivalent of W *ff* and E *f* (but is not included in the modern Welsh alphabet), and in others *Penllwyn-sarth* which, although technically 'incorrect' is nevertheless an accepted variant form in the vernacular: *Pen llwyn Sarff* 1479, *Penlloynsarph* 1602, *Penlloinsarth* 1620, *Penllwyn Sarph* 1757. The form in *-sarth* shows an assimilation of joined consonants with *ff* and *th* interchanging as they do in W *benffyg* 'loan; that which is borrowed or lent' < Lat *beneficium*, superseded by the later, and current, form *benthyg*.

There is hardly need to dwell on the W elements *pen* 'head, end of' and *llwyn* 'bush, grove, copse' for it is the final element in the full form of the name which has been the subject of enquiries. Joseph Bradney in his monumental history of Monmouthshire is confident in his interpretation, 'Penllwyn Sarph, which means the end of the serpent's grove', thus taking *-sarff* to be the W epicene n. *sarff* < Lat *serpens* (through a Vulgar Lat form) 'serpent, snake, reptile' and thus promoting an onomastic tale of doubtful authenticity.

Such an interpretation is not, of course, implausible but in conjunction with W *llwyn*, and bearing in mind the category of Welsh names like Llwyn-onn ('ash'), Llwyn-bedw ('birch'), Llwyn-helyg ('willow') etc. where the second element indicates the predominant arboreal composition of the *llwyn*, it should be noted that there is another Welsh *sarff* which is usually found in the combination *pren sarff*, E *service-tree*, a tree of the *sorbus* family, *Pyrus* (or *Sorbus*) *domestica* bearing small berries and *Pyrus* (or *Sorbus*) *torminalis* the wild service-tree, closely related to the rowan-tree or mountain-ash. The ModE *service* in this context is ultimately < OE *syrf(e)* which may in turn be derived from a form of Lat *sorbus*. In ME the pl. form was *sarves*, *serves*, with reference to the fruit (edible when fully ripe in the case of *Sorbus domestica*) which was later written as *service*, not to be confused with the common E *service* 'the condition or occupation of one who serves' etc. It would appear that W *sarff* is a fairly early borrowing from a variant form of OE *syrf(e)*. GPC notes a reference to a form *sarf* 'in the English of E. Monmouthshire' whilst, earlier, a correspondent of Edward Lhuyd *c.* 1700 noted the English form *Sarph tree* close to the Welsh border, in the vicinity of Rowlston, Herefordshire, for what was called in Welsh in that area, *Pren Sarth*.

What seems to have caught the eye of Archdeacon William Coxe at Penllwyn-sarff on his perambulation of Monmouthshire in 1800–1 was a number of sycamore trees of great size. He does not comment on the prevalence, or even the existence of service-trees, but it seems preferable to accept that the place got its name from a grove of such trees long before Coxe's day rather than indulge in a fanciful reptilian interpretation.

Penrhiw-ceibr

The strictly correct form of this name is Penrhiw'rceibr with the abbreviated form of the W def.art. '*r*, which tends to be omitted in everyday speech, preceding the final element. The first two elements of the three-element combination are W *pen*, here probably 'top, end of' + *rhiw* 'wooded slope', which, when they occur together in place-names, may be followed by a third element which often denotes the plant(s) or tree(s) to be found in abundance at that location (see p. 149–50 for Penrhiwtyn).

In his essays on the parish of Llanwynno (1878–88) Glanffrwd interpreted the third element in the present example, *ceibr*, as a reference to the thickly entwined branches of the trees through which a passage or track had to be cut to facilitate access for various means of transportation (mainly horse-drawn) up the slopes of Glyn Cynon towards the old farmhouse so named near the summit—where the village of Berthgelyn now stands—so that they looked like rafters arched over the track. Indeed, in the tithe award of 1841 another name occurs in the vicinity, namely Pantynenbren (W *pant* 'valley' + *nenbren* 'roof-beam').

Glanffrwd was not far out in his thinking, it would appear, because the n.m. *ceibr* is considered to have been adopted indirectly into Welsh through a Latin form which is the basis of the Fr *chevron* 'the beam which supports a roof', and the root form of which in the Celtic languages, including the Gaelic *cabar*, now *caber*, which is the large pole or beam which figures in the Scottish sporting activity known as 'tossing the caber'.

In its definition of the element *ceibr*, GPC has only a slight change in emphasis to the explanation offered by Glanffrwd, stating that Penrhiw-ceibr was so-named 'am fod coed addas i lunio ceibrau yno' ('because there were trees suitable for making beams there').

In the parish of Pendeulwyn (Pendoylan) in the Vale of Glamorgan there is a Coed Ffos-ceibr (with W *ffos* 'ditch, dyke') and through it runs a stream Nant Tynyplancau W *tyn* < *tyddyn* 'cot' + *plancau*, cf. E *planks*) which suggests some structural connection.

Penrhiw-ceibr is not a name of great antiquity. The earliest form attested thus far is *Rhiw'r Kibier* 1748 which shows a tendency in common parlance to introduce an intrusive vowel between the *b* and *r* of the final syllable. This has given rise to a variety of spellings over the years such as *-cibir*, *-ciber*, *-ceiber*, *-ceibir* etc. and was probably the basis for the unfortunate anglicised form of the name of the local colliery, Penrikyber, the quick demise of which, or more applicable today, perhaps, the prevention of the resurrection of which, is something to be fervently desired.

Penrhiwtyn

Locally pronounced Penrwtin, near Cwrt Sart and Ynysymaerdy, between Llansawel (Briton Ferry) and Neath, and the majority of forms of the name attested support this: *Penrwting* 1590, 1620, *Penrwtting* 1628, *Penrutting* 1631, *Penrooting* 1687, *Penruttin* 1729 etc. Its meaning remains debatable.

One of the earliest documented forms is *Penrowting(e)* 1586, 1631, *Penrouting* 1611, and on that basis, no doubt with tongue firmly in cheek, one scholar once suggested that the second element in the name was E *outing*, because the location on rising wooded ground along the banks of the river Nedd was a favourite place for the inhabitants of Neath to come to for an 'outing'. Hardly, because at least now we have the advantage of an earlier attested form of the name, *Pen-ryw-wtting*, in a will of 1558.

This allows the possibility of the name to be considered in the category of names having as their first elements the W n.f. *rhiw* 'wooded slope' (in south Wales), or *pen* 'head, end of' + *rhiw*, as in Rhiw'rperrai, Rhiwonnen, Penrhiw'rceibr, Penrhiw'rgwiail. The final element in all these is the name of the plant or tree which is to be found in abundance at the place so named—W *perrai* pear-tree', *onnen* 'ash-tree', *ceibr* 'caber, large tree', *gwiail* 'hazel (rod)'.

Consequently, bearing in mind the provection of -*d*- between vowels (> -*t*-) which is a feature of southern Welsh dialects, as in words like *wedi / weti, odw / otw, (dy)wedws / wetws* etc., a similar occurrence might have been possible in the case of the plural form *gwdyn*, sg. *gwden* 'hazel, gwiail'. This would become *gwtin* in common parlance and in a lenited form as a final element it would appear as -*wtin* in the combination *Pen-rhiw-wtin*.

In Ardudwy, Rhiw-wden is the name of the side of a mountain in a Cymer Abbey charter of 1209, and one of Edward Lhuyd's correspondents also refers to a *Bwlch rhiw wden* 1695–1709 in the parish of Llanuwchllyn, Merioneth. It is also relevant to note that W *gwden, gwdyn* occurs in an old Welsh 'botanology' as the name of that plant which intertwines the branches of other trees, in Welsh *gwden y coed*, that is, 'bindweed' or 'old man's beard'. In a wooded area, very different in its aspect from what meets the eye today, this is another possibility to be considered, particularly in view of the fact that an English naturalist in the last century commented on the thickly wooded Vale of Neath, that a feature of its woods were 'creepers hanging down in festoons'.

Pentrebane

One of the influences often misconstrued as English distortion of Welsh place-names is that which accounts for apparent changes such as Aberdare for Aberdâr, Heol-lace for Heol-las, Nant Brane and Trevrane for Nant Brân and Tre'rfrân etc. The only feature in such cases, however, which can be regarded as non-Welsh is the spelling. The sound which is presented in this way is authentically Welsh and such forms effectively preserve what was the contemporary Welsh dialectal pronunciation during the period in which they were recorded and of which, otherwise, hardly a vestige survives.

It should be realised that non-Welsh scribes, at a time when there was no orthographic standardisation, attempted to express what they heard and did so in the terms of their own language medium. Visual distortion is a relatively recent and literate phenomenon. When faced with the reality of the 'narrowing' in pronunciation of the long vowel *ā* to *ǣ* in the virtually extinct Gwentian Welsh dialect of Glamorgan as in the final syllable of Aberdâr, the nearest representation the scribe could resort to was Aberdare (to rhyme with *care*, *mare* etc,). Thus *-lace* for *-las* in Heol-las and *brane* for *brân* in Nant Brân.

In some cases the long *ā* in the Gwentian dialect was a stage in the phonetic development of the diphthong *ae*. In this sequence *ae* > *ā* > *c̄*, thus W *cae* 'field' > *cā* > *cē*, a sound for which an English scribe's equivalent was something like *kay*, this being precisely the form which occurs in scores of estate documents, plans and surveys, particularly from the sixteenth century onwards. This accounts for the form Lisvane for Llys-faen from the spoken form *L(l)ys-fān* > *Lys-fēn* > Lisvane.

Pentrebane is now the name of a Cardiff suburb, but it is originally the name of an existing farm on a 'ridge, back' of land, W *cefn*, in the parish of St Fagans. It was *Cefn-tre-baen c.* 1485–1515, *Kentrebain* 1536–9, *Keven trebayne* 1570 although the present first element pentre appears as early as *Pentrebaen* 1564, *Pentrepayne* 1567, and is firmly established as Pentrebane from about the middle of the eighteenth century, the first element having developed by analogy with the common W *pentre(f)* from *cefn-tre-* and *cen-tre-*. (For loss of *-f-* in the combination *-fn-* see p. 83). Here also we must have a similar vocalic presentation in the last syllable *-bane* to that which is noted above. The original element was the Welsh form, *Paen*, of the Norman-French personal-name *Payn(e)*, *Pain(e)*.

Pentrehaearn

This is the form of the name of a substantial farmstead on the lower slopes of Mynydd March Hywel on the eastern side of the Clydach valley (Dyffryn Clydach) on the recent Ordnance Survey Explorer map, sheet 165. The valley is that of one of eight rivers named Clydach in Glamorgan. It rises on the mountain in the vicinity of Cilybebyll and runs down to join the river Neath (Nedd) close to the site of the old Neath Abbey. On the first edition one-inch OS of 1830 the farm is named *Pentre-haiarn* and this form was retained up to the first series of the now discontinued OS Landranger maps (1982), sheet 170, then changed again to *Pentrehaearn* on that of the second series of 1992. On the earlier (and also now discontinued) 1:25,000 Pathfinder map, sheet 1107, it was *Pentreharne*.

These forms are all based on suppositions concerning the form and meaning of the two elements of which the name is composed. The first element was assumed to be the common W n.m. *pentref*, probably in the later general sense of 'village, hamlet'. The second element was understood as the W n.m. *haearn* 'iron' with *haiarn* and *harn* both acceptable colloquial variant forms, the latter being the usual form in the counties of Carms., Mon. and Glamorgan by loss of the final syllable, *haearn* > *haern* (cf. *daear* > *daer*, *gaeaf* > *gaef*) and a further reduction of the diphthong in *haern* > *harn*.

The present form of the name is based on a wrong division of the constituent elements. In the first place the second syllable of what appears to be *pentre(f)* was originally attached to *-harn* as the second element, that is *-treharn(e)*, and an examination of the earlier forms of the name reveals that the present first syllable *pen-* has been substituted for an original W *bryn* 'hill'. Among the Plas Cilybebyll papers are to be found the following forms: *Bryn(n) Traharne* 1601–5, 1671, *Brynn Treharne* 1642, *Bryntreharne* 1663–4, *Brintreharne* 1689–90, *Bryntreharn* 1711, *Bryn Treharne* 1748–9. The second element can be either the surname Traharn(e) with its variant forms Treharn(e), Trahern(e), Trehern(e) (from the W pers.n. Trahaearn, Trahaern, Traharn, i.e. the intensifying W prefix *tra-* + *haearn* in an adapted sense 'like iron, strong, resolute' as in the names Aelhaearn, Cynhaearn, Cadhaearn), or the W pers.n. itself, both used with *bryn* to signify possession or close association at one time. In view of the lack of forms to show differently it would appear that a loose pronunciation of the form *Bryntreharn* (< *Bryntraharn*) and analogy with the very common W *pentre(f)* produced *Pentreharn*, 'restored' erroneously to *Pentrehaearn* in later years.

A somewhat similar kind of treatment, with a different first element, can be seen in the name Troed-yr-harn on the A470 near Llanddew, north of Brecon. In 1580 there occurs the form *Lloyn tretyharn*, which is the W *llwyn* and a second element compounded of *tre(f)* + *Tryharn* (another variation of *Traharn*) and proof of the composition of that second element *-tretyharn* is provided by the E form *Traharneston* 1331 where the E suffix *-ton*, as is so often the case in medieval names, corresponds to W *tref*. Indeed, here the W *–tretryharn* may be a literal translation of the name of an English homestead. Further, *Tretraharn* itself was later modified to produce the modern Troed-yr-harn, the significance of which, as it stands, is otherwise questionable.

Plas Milfre

The location of this old homestead on high moorland near Cefn Brithdir, north of Bargoed in the parish of Gelligaer, is marked *Faldray* on Christopher Saxton's map of Glamorgan 1578, and those of John Speed 1610 and Johan Bleau 1645, but on Emanuel Bowen's map of 1729 it has acquired an erroneous capital letter and appears as *Taldray*.

It would appear that these forms are attempts to reproduce a name which occurs in a collection of estate documents once in the possession of John Capel Hanbury of the well-known industrial family of that name of Pont-y-pŵl, namely *Tir yffalledre* 1507–8, *Tiry Falledre* 1508, and *Tyr y Ffuldre* 1511 prefixed by W *tir* 'land' and the W def.art. *y*.

It is a form which bears a strong resemblance to the name of another old homestead close to the Roman fort near Coelbren, Brecs., east of Ystradgynlais and just over the county border in the old Breconshire, namely Tonyfildre (*Ton-y-ffildre* on some Ordnance Survey maps), *Tonvildrai* 1805 and *Tonfildra* in local usage, except that the name has the prefix W *ton* 'surface, skin' used commonly in place-names in Glamorgan, Gwent and southern Breconshire to mean land that has not had its surface 'broken' or ploughed, 'grassland, layland', rather than *tir*.

Tonyfildre can be interpreted reasonably confidently as having a compound second element *-mildre(f)*, a W n.f., its gender determined by that of the generic, W *tref* 'homestead, farmstead' and therefore having its initial consonant lenited after the W def.art. *y fildre(f)*. Its first element is W *mil* 'beast, animal' and is the element found in W animal names like *bwystfil*, *morfil*, *milgi* etc. A *mildref* might well have been a farm devoted to animal husbandry. Of the sixteenth-century forms of Plas Milfre quoted above it is obvious that they are entries made by a non-Welsh scribe and that, perhaps, it is the second element of the form *Tyr y Ffuldre* 1511 which comes nearest to an acceptable comparison with *Tonyfildre*, some liberty having been taken with the vowel quality of the first syllable of the second element and taken even further in the forms of 1507–8 and 1508 noted above. It is pretty certain that the *-e-* before *-d-* in *-yffalledre*, *y Falledre* has no value as an accented syllable and is a 'silent' *e* of an English scribe, a similar scribal confusion over the value of the initial W *f-* being responsible for the forms in *ff-*.

However, not to be ignored as a possible influence on the form of the name is that of another farm in the Clydach Uchaf valley, Resolfen, Ffald-y-dre (*Fald y Dre* 1731) at not too great a distance away down the Neath Valley from Tonyfildre. This farm dates from the beginning of the seventeenth century, perhaps earlier, and the name is of two elements, the W n.m. *ffald* 'fold, pen, pound' borrowed from OE *fald* + *tref*, 'the homestead fold', but there is no direct evidence of any analogical influence here on the forms *Tonyfildre* or *Tir y ffalledre*.

Plas Milfre was a substantial holding. The relevant volume of the Glamorgan Inventory of the Royal Commission on Ancient and Historical Monuments does not date the structural features of the building, such as post and panel partitions between rooms, diagonal cut stops and chamfered beams etc., earlier than the seventeenth century despite the available sixteenth-century forms of the name. Its later status is indicated by the acqusition of the prefix *plas*, borrowed from OFr *place* possibly through ME *plas*, *place*, in the later sense of 'country house, manor house', by the early nineteenth century. It was *Place Mildra otherwise Tye Evan Thomas* in 1834 where the *–d–* of *–fildre(f)* is still retained. The present form of the second element *milfre* is, therefore, a late substitution which is not easily explained in terms of a consonantal *d/f* colloquial change which is hardly a common occurrence. The most reasonable suggestion for this may be a popular analogy with the name *moelfre* < the W adj. *moel* 'bald, bare, round-topped' + *bre* 'hill', which is widely dispersed in Wales and also to be found in England as Mellor in Lancashire and Cheshire, and Mulfra, Mulvra in Cornwall.

Pont Llewitha

The bridge, W *pont*, carries the A4070 to Casllwchwr (Loughor) over the river now known as the Afon Llan on the northern outskirts of Swansea. River names are usually among the oldest names in existence but hitherto the earliest record seen of Llan as the name of this river is dated no earlier than 1799 on George Yates's map of Glamorgan which, in such a context, must be regarded as being comparatively recent.

On its way to the sea it joins the river Lliw, which runs down from above the Upper Lliw Reservoir to the north, near Mynydd Stafford

(Stafford Common), and it is the joint waters of both rivers which run out into the estuary of the Llwchwr. At one time, however, it would appear that both rivers were regarded as branches of one river. A reference to the estuary in the Black Book of Carmarthen's famous 'Stanzas of the Graves' states simply, and in singular terms, *yn yd a lliv yn llychwr*, 'where the Lliw flows into the Llwchwr'.

In 1306 the present Lliw was referred to as the *Norther Lyu* 'the northern Lliw' and Rice Merrick called it *Llugh ye greater* in 1587, which implies the existence of another Lliw. What becomes apparent is that it was the present river Llan which was, indeed, that other Lliw. In a charter of 1153–84 it is referred to simply as *Lyu* and, perhaps appropriately, *Llugh the lesser* by Rice Merrick.

Further evidence exists to show that other means of identifying the two rivers occur over the years. In 1697 the *Luw ycha* and *Luw issa* 'the upper and lower Lliw' are recorded, but by 1764 the present Lliw (the upper) was so named, devoid of any qualifying element, in the form *Llyw*, whilst the present river Llan (the lower Lliw) is documented as the *Llyw Ytha* (a well-known survey of Gower in 1764 notes the rivers *Llyw* and *Llyw Ytha*) a designation which has an even earlier provenance as *lliwytha* in 1600. The second element in this form, *-ytha* represents the colloquial form of the Welsh superlative adj. *eithaf* 'farthest' > *eitha* > *itha* (*ytha*) 'the farthest Lliw' (cf. *eithin* > *ithin*, p. 119), and this is the form to be seen in the present Pont Llewitha (= Pont Lliw Eithaf) 'the bridge on, or over the farthest Lliw', and recorded as *Pont Llwytha(u)* 1729, *Pont llewydde* 1833.

That this was not the only example of identifying two rivers of the same name in this way is borne out by a further reference in a sequence of *englynion* in the Black Book of Carmarthen to two rivers called *Tawue, Tawuy* (Tafwy in modern orthography, the early form of what is now Tawe), namely *y tawue nessaw* (Y Tafwy Nesaf 'the nearest Tafwy') and *y tawue eithaw* (Y Tafwy Eithaf), the former being probably a stream which rises near Llan-crwys, Carms., and the latter the Swansea Tawe.

The reason for naming the Lliw Eithaf the *Llan* from the eighteenth century onwards is not known with certainty, but its status as the principal watercourse of the parish of Llangyfelach would seem to have been a reasonable incentive to do so, when considered in the light of names such as Melin-llan, Pont-llan, and Gors-llan which exist in the parish.

Pont y Rhidyll

Perhaps it may be granted that the W n.m. and f. *rhidyll* 'sieve, riddle' seems an unlikely element to occur in a place-name with W *pont*, but, as it happens, it is the English equivalent *riddle* (the W form being a borrowing from the OE *hriddel*, ME *riddil*, *rydil*) which is the key to its meaning.

This was a small bridge, still in existence, although nowadays strengthened to cope with the weight of modern traffic, which carried the old turnpike road from Dinas Powys to Cadoxton (Barry) over the Coldbrook stream on the edge of Dinas Powys moor.

The name of the bridge is recorded in a Latin document of 1566 as *Pontem vocatur Ryddelle*, the second element varying considerably in a substantial number of estate papers thereafter—*Ridell*, *Riddill*, *Redell*, *Rydhill* and *Riddle* being amongst the most conspicuous—and it is an element which occurs as a distinguishing element in relation to the name of a well (*Riddells Well* 1576), land (*Riddell Mead* early 17c., *lande called The Redell* 1601, *The Riddle Lands* 1777 etc.) and an old farmhouse, in addition to the bridge.

It is reasonably clear that, where the farmstead is concerned, which is *The Riddle Farm* 1783–6, Riddle is the surname of its owner, probably the first, so that it would appear that the well, the land and the bridge were so named because of their association, one way or another, with the property.

By the nineteenth century, when the Welsh form *Pont y Rhidyll* is recorded, the significance of the element Riddle must have been lost, and in order to create a Welsh form it was interpreted as the English common noun *riddle* 'sieve' and translated literally as *rhidyll*.

It is pertinent to note the prevalence of the surname Riddle in Somerset in view of the fact emphasised elsewhere in these notes that the Vale of Glamorgan was subjected to a considerable influx of people from the south-western counties of England over the years. Furthermore, the property is to be found in the close proximity of two others whose names are examples of names which were originally surnames of the landholders concerned, namely Greenyard (originally the property of a certain John Grennett in the reign of Elizabeth I) and (The) Gilbert.

Pont-rhyd-y-fen

Many place-names in Wales still retain elements which indicate two ways of crossing rivers and streams where, perhaps logically, one would have been expected to give way to the other during the course of time, namely W *rhyd* 'ford' and *pont* 'bridge'. In all probability *rhyd*, which not only means 'ford' but is also cognate with the English word, is the earlier of the two in common usage. As Margaret Gelling has observed, 'probably most bridges were built to supplement the fords which were of major importance in the early years'. This is confirmed by the sequence in which the two are placed when both occur together in some names in Wales, *pont* usually being prefixed to a name which previously had *rhyd* as a generic. This is the case with Pont-rhyd-y-fen, two examples of which can be found, one on the river Tryweryn before it enters Llyn Celyn, west of the lake in Uwchmynydd, Merioneth, the other being that which is noted below, in Glamorgan. This does not preclude the possibility, even so, that in some other names the order is reversed, particularly in *Rhyd-y-bont*, a well-evidenced form in a number of parishes throughout the country but which has no element to indicate, specifically, the bridge concerned in each case.

Ryd y venn 'the wagon ford' is the form recorded in 1460–80, and *Rydyfen* 1578, as the name of the ford by the confluence of the rivers Pelenna and Afan, north-east of Port Talbot and Cwmafan, with either the W n.f. *ben* (pl. *benni*) 'cart, wagon' or its variant form, also a n.f. *men*, as the second element to indicate the kind of traffic which predominantly used the ford. When preceded by the W def.art. it is often not possible to differentiate between the two forms in a place-name because lenition of initial *b*- or *m*- in a n.f. is *f*-, thus *y fen* in both cases. In 1536–9, however, John Leland refers to *Ponte Rethven* where *pont* has been added, together with the information that the bridge was made of wood over the river Afan. This form continues in use as *Pont Rhyd y ven* 1714, the reference here also being to the bridge and not to the community which settled in the vicinity, later to take its name from the bridge.

However, before the growth of that community to village proportions it appears that the area in which it was to be contained was first designated by the prefix W *tir* 'land' in *Tir pont rydyven* 1601 and more specifically *Tyr Pen Pont Rhyd y ven* 1756 with the additional element W *pen*, here 'head of, end of' to indicate the precise location where the present village grew, both *tir* and *pen* being excluded later as elements. This shedding of elements is seen in other examples containing a somewhat similar sequence, probably as a process of abbreviation in common speech.

Pontardawe, for instance, is first recorded as *Tir penybont ardawe* 1583–4, literally 'the land at the end of the bridge on the Tawe', but *Tŷ pen y bont ar y Tawey* 1675, *Tŷ penybont ar tawe* 1710–11 where a common scribal uncertainty in documents between the W elements *tir* and *tŷ* 'house' occurs before the present Pontardawe becomes current.

Pontarddulais, without the prefixed *tir*, was *Penybont ar ddylais* 1578, and Deric John (*Notes on some place-names in and around the Bont* (Aberdare, 1999), pp. 18–23) now points to an earlier form, *Penybont aber Duleis* 1550, which raises the interesting point that if this can be confirmed as the original form, the modern form of the name may be the much maligned Pontardulais after all (with no lenition of the initial letter of the river name, *d* > *dd*, as would be the case if it were a syncopated form of that which appears in 1550). The form *Penybont ar ddylais* is recorded by the antiquary Rice Merrick but he also informs us in the same source, his celebrated *Morganiae Archaiographia*, that the 'little river' Dulais (W *du* 'black' + *glais* 'stream') 'runneth into Lochor at the bridge of Dylays', clearly implying that the bridge, at the end of which the modern settlement grew, was over the Dulais at its confluence, W *aber*, with the river Llwchwr (Loughor).

Outside Glamorgan, perhaps the best known example which developed along similar lines is Llanbedr Pont Steffan, also known as Lampeter, Cards. (this being the anglicised form of the Welsh colloquial *Llambedr* < *Llanbedr*). Although *Lampeter Pount Steune* is recorded in 1301, it is *Lampeder Talpont* 1303–4, *llanbedyr tal pont ystyvyn* 14c., *Lampeder tal pont stevyn* 1407, where *tal* 'the head of, the end of' is used, like *pen* in the earlier forms of Pontardawe above, to indicate 'the church of St Peter at the end (head) of Stephen's bridge'. From the mid sixteenth century the modern W form becomes the norm, *ll.bedr bont ystyfyn c.* 1566 etc. 'the church of St Peter at the bridge of Stephen'.

Radur

On 10 June 1944, a sunny summer afternoon, a few days after the Normandy landings by the Allied armies, a large detachment of German troops and tanks entered the rural village of Oradour sur Glane near Limoges in south-western France. There, 642 of the inhabitants, men, women and children, were burnt to death or shot and every building ruinated. Ever since, Oradour is one of those names, like Lidice and Lezaky, which resound horribly in the mind. The ruins of the village still remain abandoned, just as the German troops left them. It is an eerie and unforgettable experience to visit the place.

We now know that the name Oradour is an early form 'fossilised' in its development from the Latin *oratorium* 'prayer house, oratory, chapel'. It occurs elsewhere in France in forms which developed later in French, like Ourouer, Ouvrouer, Ouzouer etc.

Another form which can be derived from the Latin *oratorium* is that which occurs twice as a place-name in Wales, namely Radur (not Radyr) near Cardiff and another Radur (*the chapel of Radour* 1556–8) near Usk in Gwent, despite the efforts of those who imposed the image of a plough, W *aradr*, on the wall of the Radur Comprehensive School in an effort to explain the name.

The form *radur* is the colloquial form of the basic *arádur* (by loss of the unaccented first syllable) from the Vulgar Latin form *ŏratorium*, with its short initial vowel, a development which can be compared with that of W *achos* < *ŏccasio* or W *achub* < *ŏccupo*. *Aradur* is the form in the *Liber Landavensis* and is found documented until the sixteenth century: *Aradur* 1506, *Aradyr* 1533, *Aradier* 1554 etc.

Since the first syllable was unaccented, *arádur*, and before it was lost in common parlance, the belief arose that the unclear sound was the vocalic form of the W def.art. *y*, so that forms like *Yradur c.* 1569, *Y Rader* 1587, *Yr Radyr* 1592 and *Yr Adyre* 17c. make their appearance to be imitated in an English version *the Radyr* 1554, *the Rader* 1583, *ye Radour* 1675 and *the Rader* on the Glamorgan maps of Saxton 1578 and John Speed 1610. There is no justification for the use of the def.art. in Welsh or in English.

How different the experience of Oradour and Radur. One notices that as part of the present vogue of 'twinning' towns and villages with others abroad the coupling of these two villages does not seem to have been contemplated. The common derivation of their names could well be added to the desire to commemorate a great act of human sacrifice as a motive for doing so.

Radyr Chain

This is the current form of the name of this area surrounding a crossroads on the A4119 between Llandaf and Radur. It has given estate and land agents an opportunity to adorn their property advertisements with the modish spelling Radyr Cheyne in an effort to bestow upon it some of the grandiloquence suggested by the occurrence of *cheyne* (more correctly *chine*) as an element in fashionable addresses on the south coast of England, in places like Bournemouth. Of course, in that part of the world *chine* does have a meaning. It comes from OE *cinu* 'narrow valley, glen', but this does not apply to Radyr Chain.

In contrast, what we have here is the common English word *chain*, and the reference is to a chain placed across the narrow road or track which led down from the crossroads, past the small church of Radur and Radur Court, over the river Taf, and on to the old Coalyard of Llandaf (broadly the area where the present Glan-taf High School stands).

The road between Llandaf and Llantrisant was a turnpike road and in the nineteenth century the trusts who ran the turnpike system erected or placed a barrier, in this case a chain, across any minor road that branched off the main road, especially if the minor road was likely to be in frequent use.

After the Rebecca Riots in the early nineteenth century when the destruction of tollgates on turnpike roads was the characteristic activity of Rebecca's 'daughters', a Royal Commission looked into the causes of the disturbances in 1844, and this is part of the answer of one witness who was asked about the chain on the road near Radur:

> It is the Radyr Chain: it is to prevent people going towards Llandaff coal-yard principally, turning off from the turn-pike road into the parish road without paying.

Scarcely could the situation have been put more succinctly. The chain was sufficiently well known to give the crossroads its name, duly etched on the first edition of the one-inch Ordnance Survey map in 1833.

Rasau

Although not included in the old county of Glamorgan, the upland settlement of this name in the vicinity of Beaufort and Ebbw Vale is sufficiently close to its north-eastern boundary to justify its inclusion in this volume. It is included mainly in response to many enquiries regarding its form and meaning for it is all too often subject to unfortunate treatment under English influence which leads to its mispronunciation and incorrect spelling, particularly on maps. This arises from the careless habit of doubling consonants in accented syllables in Welsh names which are not so treated in the Welsh language. Betws sometimes appears as *Bettws*, *Sketty* for Sgeti, *Kilgetty* for Cilgeti etc., but there is no *-tt-* in Welsh. *Llandyssul* is often written for Llandysul, *Abertysswg* for Abertyswg etc. where *-ss-* is incorrect. Likewise Rasau is too often seen as *Rassau* with sometimes comical results when it is frequently pronounced as if to rhyme with the name of the German town of Nassau, or that of the capital of the Bahamas. The Welsh ending *-au* is not pronounced as *–aw* in *law*, *saw*, etc. but as in Welsh words like *dau* 'two', *cnau* 'nuts', *parhau* 'to last' etc. and is in this particular instance the common Welsh plural ending *-au*.

Rasau, colloquially *Rasa*, is the simple Welsh plural form of *ras*, borrowed from E *race* in two senses, namely that which is run, a contest of speed and, in the present case, as a place-name element meaning 'a flow, a channel of water', as in the compound *mill-race*. The term goes back to the early days of mineral extraction when mining processes were relatively primitive on the 'patches'. It was a small-scale type of opencast working of shallow levels of iron ore in particular, coal subsequently, which was mined and deposited in locations above which ponds were excavated and their walls breached to allow a flow of water to run down, so to wash or 'scour' the ore of adhering soil and impurities. This free flow of water was later directed along a constructed watercourse. This was the *race*, as is recalled at the head of the Rhymni valley in *a gully named Brynoer race* 1816 which was, in its Welsh form, *Ras Bryn Oer* 1832. It would seem that natural watercourses were also used for the purpose where they were conveniently located as is borne out by the existence of several surviving examples of the name Scouring Brook, such as that on the northern slopes of Caerffili mountain, and in the Welsh form Nant y Sgwrfa near Aberpergwm in the Vale of Neath (*Nant Yscwrfa* 1884, another Welsh borrowing based on the stem of the verb *sgwrio* 'to scour' + *-fa* 'place').

The historian of Ebbw Vale, Arthur Gray-Jones, reminds us that in the early days of mineral exploitation in the area, during the seventeenth century, the few small iron furnaces which existed in the valley of the Ebwy obtained their ore from sources which included the southern slopes of Llangynidr mountain in the vicinity of Rasau. Indeed, the area was noted for its 'scourings' and races, we are told, and it would appear that originally, when the name was first given, it was to the area as such, as the location of many races as it were, that it alluded. Confirmation of this comes from the leases of 1779–1801 which Edward and Jonathan Kendall obtained from the duke of Beaufort to establish their ironworks at Beaufort (*y Cendl* to the Welsh inhabitants, being a cymricised version of Kendall) and to work the veins of coal, iron ore and limestone in the area. These refer to *the Rhassa*, the def.art. making the location a specific one which was later to be retained as the name of the modern settlement which grew there. However, the name must have been in existence before the date of the documentation which suggests that in an area which has since seen a marked decline in the use of Welsh in daily converse the language was sufficiently prevalent in common parlance in the eighteenth century to ensure that it was the Welsh borrowing of an original English term, not the English term itself, which was applied to identify an area by reference to its main features.

The term also survives in the names of various other places affected by early industrial activity of a similar nature. Near Pontypŵl there was *the Race* 1698, after which *Race Farm c.* 1735 was named, now fragmented into two holdings, Upper and Lower Race. In the parish of Llangyndeyrn, Carms., the term exists in names related to similar activity in the Gwendraeth Fach valley: the present day Rhas Cottage, Pen-rhas, Pant y Rhas, being connected with *The Ras* 1726, and Rhas Fach which was *Rhace vach c.* 1695, *Rhas* 1831. Also the large pond still in existence on Twyn-y-waun above Merthyr Tudful bears the name Rhaslas or Pwll Rhaslas (with the W adj. *glas* which has a range of meanings from 'blue, greenish blue' to 'grey, clear, translucent') which was probably a relic of the scouring process.

One word of caution is necessary in the event of the word *ras* appearing in a place-name form in Welsh manuscript sources such as the Peniarth MSS in the National Library of Wales. In pedigrees in volumes dated *c.* 1545, *c.* 1566 and *c.* 1569 respectively the forms *yras fach*, *y ras*, and *plas y ras* refer to Nash Manor or Little Nash, between Cowbridge and Llantwit Major. In these forms there is a wrong division of letters. The name Nash in this case is derived from the ME form *atten ashe* 'at the ash tree' with *ash(e)* borrowed into Welsh as *as*, with the def.art. *yr as* (not *y ras*).

Rhoose : Y Rhws

As is well known, Cardiff International Airport, the major portion of which lies in the parish of Pen-marc (Penmark) in the Vale of Glamorgan, is located on the outskirts of the sizeable village which bears this name and is itself in the parish of Porth-ceri (Porthkerry). In the sixteenth century there stood in the vicinity a house of the Mathew family which was of sufficient status to be designated *Plas Rws c.* 1569 and *Y Rwss* 1540–77 which became *Yr Hws* by 1606–20 under the erroneous impression that the initial *r-* of *rws* was the attached final *-r* of the full form of the W def.art. *yr* and probably by analogy with *hws*, a W borrowing from E *house*, OE *hus*.

However, there is little doubt that the basic form of the name is the W n.f. *rhos*, the earliest form evidenced to date being *Rhos* 1533–8, but before the end of the sixteenth century the forms *Rhoose*, *Rouse* 1538–44, *Rowse* 1587, *Roose* 1595–6, *Roouse* 1596 etc. are recorded.

The meaning of W *rhos* in place-names varies according to location. Although it has now acquired a general sense of 'moorland', examples occur of its use in the sense of 'moor, heath' or 'high promontory, peninsula', even 'rough upland grazing'. It derives from a British form *rosta, PrW ros, which seems to have been borrowed into English at an early date to produce the form Ross, in Herefordshire and Northumberland. It is cognate with Ir ross 'promontory, wood', Bret ros 'hill', Corn *ros 'upland, high ground' also 'promontory' in some cases. The present village of Rhoose is some distance from the sea shore and stands on ground which is about 200 feet above sea level, so that the sense 'moor, heath' may well be appropriate to its location.

The form taken by the name is interesting. It would seem that an early variant form, PrW rōs (with a long vowel) must have evolved, which Professor Kenneth Jackson has suggested was borrowed as an Anglo-Saxon *rōs which produced forms such as those seen in Roose, Lancashire and Roos in Humberside. Since there is otherwise a complete lack of very early forms for the Glamorgan Roose, the -oo- looks like a later analogical rendering of the -ō- in W rhos during a known period of over-rounding of OE -ō- to -o(u)-. This is probably true of Roose in Pembrokeshire and it is relevant to note the fourteenth-century forms *Roos Ysdulas* 1335 and *Eglwys Roos* 1358–61 for Rhos Is Dulas, Conwy.

The Rhyddings

As on the 1843 tithe map of St Mary's parish, Swansea, this appears to be the accepted modern spelling of the name of this once substantial residence on the ridge above the present St Helen's rugby and cricket ground. It is also incorporated in the present street-names Rhyddings Park Road and Rhyddings Terrace.

The eighteenth-century house may have been raised on an earlier site and was the house of Thomas Bowdler (1754–1825), he whose surname became the basis of the verb 'to bowdlerise' in reference to the process of removing words or phrases regarded as indelicate in classic literary works and who produced his expurgated edition of Shakespeare in 1818.

The original holding was named after the land on which it stood, the form being based on the OE term *ryding* 'a clearing', a fairly common term which survives either as a field name or as an element in minor place-names in Wales as well as in England. In many examples in England it can denote 'an assart; land cleared of wood and scrub' or, sometimes 'land taken into an estate from waste', being possibly based on an unrecorded OE verb *ryddan 'to rid, to clear'.

Over the years the form varies in documentation and can occur as *Rydings, Riddings, Reedings* and *Reading* among others. It is recorded near Oystermouth and Cilibion in Gower, near Cadoxton-juxta-Neath (Llangatwg Nedd) and as the name of Reddings Farm near Tintern Abbey in Gwent. In north Wales it is also recorded in Denbighshire and Flintshire, about thirty examples in east Flintshire having been found by Dr Hywel W. Owen.

The Swansea example cited here is first evidenced in a document of 1402 which records a transfer of land to the Stradling family in or on *les Redyngez*, followed by *Reeding* 1617, *the Riddinges* 1650 etc. the final *-s* being, seemingly, a common colloquial accretion in English field names in the Vale of Glamorgan which does not appear to have any real syntactical significance as an indication of a plural form and having possibly originated by indiscriminate analogy with forms in which it has a genitival significance, as in *breeches*, *splotts*, *bottoms*, *little yellowes*, *downs* etc.

The form Rhydding, near Cadoxton-juxta-Neath (*Rhyding* 1626–7, *Rheeding(s)* 1661, 1666, *Redding* 1676) can be accepted in a similar sense as indicating a later stage of clearing the surrounding wooded area initiated by the monks of Neath Abbey, such as is indicated by the name of one of the abbey's granges, Cwrt Sart, near Ynysymaerdy, the final element *sart* being a syncopated colloquial form of the ME term *assart* noted above and which came through French < Latin *exartium*.

Rhydfelen

Is it Rhydfelen or Rhyd(y)felin? Many enquiries have been received concerning the name of this populous village on the southern fringe of Pontypridd, especially since both forms, by some mischance, appear on signposts on the A470 and despite the naming of the local Welsh-medium secondary school as Ysgol Rhydfelen.

That the first element refers to a ford, W n.f. *rhyd*, on the river Taf is highly probable but the balance of evidence prompts the acceptance of the W adj. *melyn* as the second element, in its feminine form *melen* to agree with the noun, and lenited initially for the same reason, rather than the W n.f. *melin* 'mill' as espoused by the mapping fraternity and others. Rhyd-y-felin is a very common name in Wales, and it appears to mark this location on the 1885 edition of the six-inch Ordnance Survey map. As far as is known at present, there is no evidence of the existence of a mill (probably of the water grist variety) in this vicinity, which goes back to the sixteenth century, as noted below. The claim made in a recent publication that there was a water-driven corn mill on the east bank of the Taf near the Dyffryn at Rhydfelen which 'was gutted one night in a spectacular mass of flames and was not rebuilt' is not supported by firm dates. Other mills mentioned in the same context are all of nineteenth-century vintage.

The wide range of meaning which Welsh colour adjectives often have is well illustrated here because the water running over a well-used ford was probably discoloured or of a yellowish-brown hue as suggested by the adj. *melyn*. The range of colour conveyed by *melyn* extends from 'yellow, golden' to 'light-bay (of a horse), sallow or brown'. The pl. form *melynion* appears in *Rhydau Melynion*, *Rhydie melynion* 1584–5 in the parish of Llangeinwyr (Llangeinor). Another W adj. having the same kind of application is *coch* 'red, reddish brown' as in *Rhyd Goch* 1666, in the Rhondda valley, or *Rethgough* 1536–9, *Rhyd Goch* 1614, *Rhyd Goch upon Clydache* 1638, Llantrisant. Where colour is not indicated, a muddy, churned up condition is often denoted by another adj. such as the E *foul*, OE *fūl* in a name like Fulford in Somerset, Yorkshire and Staffordshire, cf. the W *halog* 'muddy, dirty, polluted' (p. 56).

The survey of the earl of Pembroke's land in Glamorgan in 1570 refers to *ynys ryd velyn* in this area (where *ynys* here probably means 'waterside meadowland') the English scribe being not too knowledgeable about Welsh feminine adjectival forms, but documents of the Bute estate from 1630 onwards refer to *tir rhyd velen* 'the land of Rhydfelen', and it was a school at Rhydfelen (with a population of more than a thousand at the time) that William Roberts (Nefydd) visited as the South Wales Agent of the British and Foreign Schools Society in 1856. William Thomas (*Glanffrwd*), in his essays on the parish of Llanwynno, comments on the appearance of the valley of the Taf which he looks down upon from the top of Graig-wen, Pontypridd as extending *heibio i Drefforest, a'r Rhydfelen, Nantgarw, Tongwynlais a Chaerdydd* ('beyond Trefforest and Rhydfelen, Nantgarw, Tongwynlais and Cardiff').

A further example of this name with the elements in reverse order is Y Felenrhyd near Maentwrog, Merioneth, *y Uelen ryt* in the fourth branch of the *Mabinogi*, which became *Lenthryd* in common parlance.

Rhydlafar

The well-known orthopaedic hospital, officially the Prince of Wales Hospital, but more familiarly known to people in Cardiff and its environs simply as Rhydlafar, is now but a memory. The establishment has closed down and the site, remarkable for being what was virtually a collection of interlinked Nissen huts, is now rapidly being built over to create an urban sprawl on the western outskirts of the city. Originally established as a hospital for the United States Air Force in the Second World War, it was adapted as an orthopaedic unit in 1953 and took its popular name from that of a farmhouse which stands opposite the site on the southern side of the A4119 road from Cardiff to Llantrisant and in the parish of St Fagans.

Rhydlafar was a Welsh gentry house of sufficient standing to be called a 'mansion house' in 1666, and it acquired its name because of its proximity to a ford, W *rhyd* 'ford', on a stream which rises on the marshy ground of Gwernycegin (pp. 89–90) further up the ridge at the foot of which the present house is located. The stream flows under the present A4119 and across *Rhydlaver Moor* 1735 to join the Dowlais brook which has its confluence with the river Eláí (Ely) to the north-west of the village of St Fagans. It should be noted that although Rhydlafar stands by the roadside at present, the forerunner of the A4119 was a turnpike road which did not come into existence until the second half of the eighteenth century. J. Barry Davies reminds us that William Thomas, the diarist of Michaelston super Ely, records on 8 May 1767 that he had measured land in the vicinity 'for the turnpike road'.

The name is well documented but possibly the earliest forms recorded are not as early as might have been expected for genealogical sources imply the existence of the house in the early decades of the fifteenth century, possibly a little earlier, and in 1503. The Royal Commission on Ancient and Historical Monuments also classes it on structural evidence as having been originally a medieval house reconstructed in the seventeenth century. John Leland notes *Lewys Lluelen a mene man of land at Reth lauar* in 1536–9, and thereafter the following forms occur: *Rydelaver* 1570, *Ryd lauar* 1578, 1626, *Ryde Lavar* 1579, *Rhydlavar* 1604, *Reedlavar* 1620, *Ryd lavarre* 1621, *Redlaver*, *Rhyd lavar* 1625, *Ridlavar* 1620, 1626, *Ryd lavar* 1630, 1638 etc., with hardly any substantial variation to cause a revision of the opinion that the second element is anything other than the W adj. *llafar* 'talkative, vocal', probably in the applied sense of 'resonant, sonorous' etc. like Tennyson's brook, cognate with Ir *labar* 'talkative', Bret *lavar*, MCo *lauar*, Celt **labaro-*.

The original location, therefore, was on or near a ford on the stream noted above, the only uncertainty arises in determining whether *llafar* in this case is the name of the stream or whether it is the qualifying adjective used to describe a particular feature or attribute of the ford, namely its shallow 'noisy, babbling, resonant' nature. R.J. Thomas, the eminent authority on Welsh river names, believed that it was the name of the stream. He noted the existence of several river and stream names in Britain and Europe whose names are based on the Celtic root form **labaro-*, such as Laver, Levern, Lowran in Scotland, Laver in Yorkshire, Laber in Bavaria and Gaul and two other examples of Llafar in Wales, one running into Llyn Tegid (the Bala lake) near Glan-y-llyn on the boundary of the parishes of Llanuwchllyn and Llanycil, Merioneth, and the other a tributary of the Ogwen near Gerlan, Bethesda, Caerns. However, it is pertinent to note that there is no record of the stream being specifically referred to by this name in this case. Further, in Welsh place-names which contain *rhyd* + a river name, the earlier Welsh syntax is *rhyd* + the preposition *ar* 'on' + the river-name, but neither is there any record of this form available. Indeed, the only direct reference to the stream is as *Nant Rhydlafar* 'the Rhydlafar stream' late in the nineteenth century, as on the Ordnance Survey six-inch map of 1884.

Rhydybilwg

The Llanishen (Llanisien) reservoir on the northern outskirts of Cardiff receives the water of several streams, none of which merits the label of 'river', not even that which is called Nant Fawr (W *mawr* 'big, great') and which runs down to the reservoir from the ridge of Craig Llanisien only to emerge and to run down to Roath Park lake and onwards, under the name of Roath Brook, into the Rhymni to the east of the city.

The area was open agricultural land before the urban expansion of Cardiff northwards in the latter half of the twentieth century. Some scattered farmsteads survive, their land still not overbuilt, with indications of former rural activity retained as elements in their names, particularly those which are associated with, or stood close to natural or man-made features sometimes no longer recognisable. Still later, in modern times, such names are adopted as area or district names.

No term is more evocative of a busy rural past than the W n.f. *rhyd* 'ford' which marked the crossing of rivers and streams, one of the most important early features in early cross-country communications. The streams in this area were hardly broad enough to merit the wholesale erection of bridges, and those that were built were of the small wooden variety, mainly footbridges, as illustrated by the name of the farm of Pontprennau (*Pontprenny* 1735–85), W *pont* 'bridge' + the pl. form of *pren* 'wood, wooden', now the name of the district being rapidly urbanised to the north-east of Cardiff.

As Professor Melville Richards has pointed out, W *rhyd* occurs as a generic in place-names with the widest possible variety of qualifying elements but from among many which are listed by him one which is missing is that which is an allusion to the shape or form either of the ford itself or that of the stream at the point at which it was forded. The name Rhydybilwg falls into this category, being the name of a ford on a small tributary of the Nant Fawr mentioned above and which became the name of a farm close to the present Llanisien reservoir. Its name is attested in local land surveys dating from the seventeenth century: *Rhud-y-bilooke* 1650, *Rhyd y billwg* 1653, *Rhyd y Billwhe*, *Rhyd-y-Byllwch* 1702, *Rhyd-y-bilwg* 1833, where the second element *bilwg* would appear to be a Welsh borrowing, or more specifically a Welsh enunciation of E *bill-hook*, that tool which has a curved or hook-ended blade which is used for hedging, pruning and cutting brushwood and the like. Thomas Richards of Coychurch has 'bilwg, *an hedging bill*' in his Welsh–English dictionary of 1753, and interestingly enough the OED quotes a reference in 1611 which defines the word as 'a Welshe hooke, or hedging bill made with a hooke at the end ... we call it a Bill-hooke'. This suggests that the ford was on or near a sharply curved bend in the stream.

What strengthens this belief is that by the middle of the nineteenth century what appears to be a 'new' English translation of Rhydybilwg makes its appearance, particularly in voters' lists and local government documents, as in *Hackerford Farm* 1846, with the popular English term *hacker* for *bilwg* 'billhook' used with E *ford*. This form persists in Hackerford Road, on the Hackerford Estate in the area developed in the latter half of the twentieth century.

Rhydygwreiddyn

Another example of W *rhyd* 'ford' in a place-name which has been the subject of a number of enquiries is that of this old farmstead in the northern hilly region of the parish of Llanwynno. 'A hump-backed old dwelling ... under a heavy load of thatch', says Glanffrwd in 1888. Rhydygwreiddyn is compounded of W *rhyd* + the W def.art. *y* + an element which is a term used for a natural feature to indicate the location of the ford. However, it is a term which is not now so easily recognised in the sense in which it occurs in the place-name and is not prominently featured in such a sense in dictionaries. This is the W n.m. *gwreiddyn*, a diminutive form in *-yn* of *gwraidd* 'root' but occurring here in the figurative sense of 'a point of origin, source' with particular reference to a spring of water: 'ford by (of) the spring'.

The name is recorded as follows: *Tir Ryde y gwrythen, Tire Reyde y gwrwythin* 1594, *tir Rhyd y gwreithyn* 17c., *Ryd gwrythin* 1630, *Tire Ryd y Gwrithyn* 1638, *Rhyd y gwreiddyn* 1671 etc. The spring is one of several which feed the turbulent brook now called simply *(y) Ffrwd*, that is the W n.m.and f. *ffrwd* 'swift stream, torrent' but whose older name was *Frutsanant c.* 1147 (= *Ffrwdsanan(t)* having an early Welsh female personal name as second element, in all probability) and being the northern boundary of land granted to Pendar, an abortive house of Margam Abbey, in the middle of the twelfth century.

The use of *gwreiddyn* in this sense is evidenced elsewhere in north and south Wales as well as in Glamorgan. A farm in the parish of St Nicholas in the vicinity of which several rills rise and run into the river Ely (Elái) is named Gwreiddyn (*Gwreiddin* 1722, *Gwrithin* 1784, *Grythin* 1824, *Gwryddyn* 1840). In the parish of Merthyr Mawr (see p. 128), with W *pwll* ' pool, pond', there is a Pwllygwreiddyn near the confluence of the rivers Ewenni and Ogwr, but perhaps the most attractive combination is in a form of which there are two examples, namely Bachygwreiddyn, with W *bach*, that element which is to be seen in W *cilfach* 'nook, corner, bend'. One farm of that name stands close to the M4 motorway near Pont-lliw and on a bend of a tributary of the river Lliw (*Bach y gwryddyn* 1584–5), the other in the northern extremity of the county on Nant-y-bryn near Coelbren (*Bach y greyddin* 1583, *Bachygwyryddin* 1729).

Rhydypennau

Yet another name in W *rhyd* in the northern suburbs of Cardiff which appears to have as its second element the pl. form *pennau*, sg. *pen* 'head, end of' etc. in most contexts but also used in attributive senses, one being *pen* 'source, spring' and on the face of it a parallel to Rhydygwreiddyn. As noted in that connection, the area to the north of Cardiff before the housing expansion of the twentieth century was open countryside lying south of the ridge which extended from east to west, of which Craig Llanishen formed a part, to form the northern limit of the old commote of Cibwr which became the Norman lordship of Cardiff and from which there ran southwards a number of watercourses, in particular the Nant Fawr.

However, the pl. form *pennau* in this case, if it is taken to mean 'springs, sources' presents a difficulty in that it is hardly feasible to accept the notion of a ford on more than one watercourse. Furthermore, some estate maps place the old farmhouse of Rhydypennau, (so-named after the ford) on the banks of the Nant Fawr and a considerable distance from its source on the northern ridge referred to above, indeed, at a point lower than its emergence from the present Llanishen Reservoir, close to where it is crossed by the present Rhydypennau Road. This latter location may well have been the approximate site of the ford.

In view of this, it may well be appropriate to note what Sir Ifor Williams had to say well over seventy years ago concerning one particular sense in which the word *pen* may occur in place-names, possibly having had in mind another, perhaps better known, example of Rhydypennau as the name of a village on the A487 road near Bow Street, Cards., for there is hardly any significant variation in the collected forms of the Cardiff name to aid interpretation: *Reed y Penny* 1767, *Reed a penne* 1773, *Rhydypenna* 1789, *Rhyd y Penna* 1824 etc.

In the agricultural world, says Sir Ifor, a farmer's stock of cattle is always referred to as so many *heads* of cattle ('hyn a hyn o *bennau* sydd gan ffarmwr, sef o wartheg a bustych yn ei stoc'), so that it is possible that it is in this synecdochic sense that *pennau* should be understood in Rhydypennau, 'a ford used by cattle, the cattle ford' cf. the W *ychen* 'oxen' used in Rhydychen, the Welsh name for Oxford.

Roath

The earliest recorded place-names in Glamorgan are those of Romano-British forts and settlements noted as being in existence in the area in Ptolemy's *Geography*, composed in Greek about AD 300, and the *Cosmography* compiled by an anonymous cleric of Ravenna *c*. AD 670, its British section being based on a fourth-century source but dated 'soon after AD 700' by a reliable authority. Many of these place-names contain river names because a large number of Romano-British forts and settlements were named from the rivers on which they were located and can be regarded basically, therefore, as topographical statements which became habitative names.

One example only is noted in Ptolemy's *Geography*, namely (in its latinised form) *Ratostathybius*, or as now preferred, the variant form *Ratostabius*, a difficult name which appears to be a river name placed by Ptolemy between the rivers Tywi (*Tubius*) and Severn (*Sabrina*). Since no other name in Glamorgan is accounted for, one recent view tends to relate the second element *-tabius* to a British **Tabios* < **Taba*, as a possible origin of the name of the river Taf + the derivational suffix *-io*, but ignoring the effect of ultimate *i*-affection, the name being tentatively identified as bearing some relationship to that river, if not a direct reference to the river itself. However, the possibility that the name of the river Taf derives from a Br **Taba* is no longer acceptable. It is more probable that a root-form **Tam-* is the source, for both Br *-b-* and *-m-* by lenition would give W *-f-* (see p. 25).

The *Ravenna Cosmography*, noted above, has an entry in its list of rivers which occurs between the Usk (*Isca*) and Ewenni (*Aventio*) which would seem to determine its broad geographical area. This is *Tamion*, which is taken to be Br **Tam-i̯o-*, the suffix *-i̯o* being often used to form names of forts and settlements derived from river names although several river names themselves are also formed with it so that this is not conclusive evidence as to whether the Ravenna *Tamion* is a river name or that of a fort or settlement, despite its position in the list. A.L.F. Rivet and Colin Smith, the authorities on Romano-British place nomenclature, can only suggest that Ptolemy's *Ratostabius* may incorporate a river name, its whereabouts unknown, and if the Ravenna *Tamion* was a fort it could have been, after all, one of those which preceded the known late Roman stone fort at Cardiff.

Another difficult aspect of interpretation is the existence of the first element *Rato-* (in which case the medial *–s-* is considered to form no part of either the first or second element but that it exists in the name merely as 'a composition consonant'). This Ptolemaic form has been the subject of intense speculation for years, one reason being the existence of the place-name Roath in the Cardiff area which bears a superficial resemblance to the first element. Any direct connection with Roath is unlikely, but it is admitted that Ptolemy's *Rato-* may well be related to a British feminine **rātis* 'earthen rampart, fortification' which gave W *rhawd* in compounds such as *bedd + rhawd*, ModW *beddrod* 'grave-mound, tomb, tumulus', and *rhawd + gŵydd*, ModW *rhodwydd* of similar meaning. At the same time, if the first element of *Ratostabius* is related to Br **rātis* 'fortification', it is difficult to accept it as an element in a river name, if such it is, because forts were usually named from rivers and not vice versa. Furthermore, it is now being pointed out that there existed a Br n.m. **ratis* 'fern, bracken' which was the root-form of W *rhedyn* < **ratino*, Ir *raith*, OCo *reden*, of similar meaning, but again, possibly, not a very likely candidate for use as part of a river name. Any connection with, or reference to the Roman fort at Cardiff is also discounted.

Cognate with the W *rhawd* is the Ir *ráth* 'fortification', from which the W *rhath* may be a borrowing and concerning which GPC goes as far as to say (in translation) 'This could be the word which is to be seen in Y Rhath, Cardiff'. This is the form recorded in early sources for Roath: *Rahat, Raath* 13c., 1306–7, 1314–15, 1321, 1351, *Rothe* 1549 etc. and *Y Rhath* in later Welsh sources, *Y Raff* (for *Y R(h)ath*) *c.* 1566 etc., but evidenced at no earlier date, as far as is known. One possible alternative is noted, after Sir Ifor Williams, namely that it could have a connection with the stem of the W verb *rhathu* 'to scrub (off), scrape, file (down)' but only, presumably, as an element in the stream name *Rath* which occurs in the place-name Amroth, Pembs. (which is *glann rath*, *Amrath* in the *Liber Lan-*

davensis) and is acceptable as the name of a stream that 'scrapes' its banks but is not relevant in this context. The incidence of the form *rath* either as a simplex place-name form or compounded with another element is well known in Pembrokeshire. Dr B.G. Charles notes the various forms in his study of Pembrokeshire place-names but, with the exception of Amroth, offers no more than a comparison with Ir *ráth* 'fort' and W *rhath* 'earthwork, fortification, mound' (*rath* being a local dialect form) concerning its etymology.

If the name is of Welsh origin, and its earlier dated forms listed above favour *R(h)ath*, it is as well to note that in the medieval manor of Roath the old manor house was surrounded by a *fossatum* or ditch. Rice Merrick, in 1578, speaks of a mound or 'pile' there: 'Within it stood an old Pyle, compassed with a Mote, which is called "the Court"; but now in ruyne'. No surface evidence of the site of the moated manor house exists, but the Royal Commission on Ancient Monuments accepts that it may well have been that of the present Roath Court, a nineteenth-century mansion south-west of St Margaret's Church, Roath.

Ruperra

What has been called the 'castellated mansion' of Ruperra stands below the eastern extremity of the ridge that runs east and west of Craig Llanishen and Cefn Onn some four miles east of Caerffili. It continues to be the subject of debate between property owners on the one hand and conservationists on the other. Although in its present form and condition it may not appeal to all tastes—a square shell with towers at each corner—it has a certain ambience which has not entirely lost its appeal. Raised *c.* 1622–6 for Sir Thomas Morgan, steward of the household of the earl of Pembroke at the time, on a site possibly established by a branch of the Lewis family of Y Fan, Caerffili, since at least the middle of the fifteenth century, it was one of what the Royal Commission on Ancient and Historical Monuments calls 'make-believe castles' popular in Elizabethan and Jacobean times based on 'designs derived from late medieval military architecture but translated into Renaissance terms'. A seventeenth-century drawing shows the 'castle', as it was termed, in a more attractive guise with gabled dormers along the walls rather than the severe embattled parapets of the present structure, which is a reconstruction following a fire in 1785. Its present condition is the consequence of another disastrous fire in 1942 which destroyed the interior, leaving the main structural walls only.

The earliest recorded forms of the name are as follows: *rriw r perre* 1550, *Rhiw'r perrai* 1560–1, *Rywrpperrey* 1572, *Ruperrey* 1578, 1581, *Rewperie* 1596, *Ruerperry* 1612, and on to *Ruperra* 1644–6 etc., all of which point firmly to a combination of the Welsh elements *rhiw* '(steep) slope, hill(side)' + the affixed form of the W def.art. *'r*, subsequently to be lost, + a substantive form *perai*. In such a combination the terminal *-ai* (as also happens generally with the pl. ending *-au*) is reduced to *-a* in common speech, thus producing the form *Rhiwperra*, with the *-r-* being doubled under the strong accent on the penultimate syllable. Further, the form of the leading element *rhiw* often becomes *ru-*, *rhu-*, under English influence as in *Ruabon* for Rhiwabon, Denbs., *Ruderin* for Rhiwderyn, Mon., *Rua* 1787, *Rhua* 1819, for the pl. form Rhiwau in the parish of Gwenfô (Wenvoe), to give the form Ruperra.

The final element *perai* is defined in GPC as a borrowing from ME *pereye* (OFr *pere*), ModE *perry*, the drink made from the juice of pears, but in this instance, taking into consideration the nearby thickly wooded hillside named Craig Ruperra on the modern Ordnance Survey map, with its wood, Coed Craig Ruperra, it seems more reasonable to think of the other *perry*, OE *pyrige*, *pirige* 'a pear-tree' which is very well evidenced in English place-names and would thus give the meaning 'hill of (the) pear-trees'.

The ridge in question is one that has seen human activity since an early date. It is topped by a multivallate Iron Age hill-fort inside the ramparts of which is the site of a later motte, possibly of Norman origin. Whether the slope is, in fact, the *rhiw* of Rhiw'rperrai cannot be established with certainty but the name conforms in character with other Welsh place-names which have *rhiw* + the name of a plant or tree such as Rhiwfallen (W *afallen* ' apple-tree'), Rhiwonnen (W *onnen* 'ash-tree'), Rhiwgriafol (W *criafol(en)* 'rowan-tree') and, preceded by *pen* 'head', Penrhiw'rceibr (p. 148), Penrhiw'rgwiail (W *gwiail* 'with, sapling') etc.

Schwyll

On the side of the road from Ewenni to Aberogwr (otherwise known by the unimaginative and not very inspiring name, Ogmore-by-Sea) stands the Schwyll Water Pumping Station of Welsh Water, first erected by the old Mid Glamorgan Water Authority, itself a successor of the earlier Bridgend Gas & Water Company, established in 1869.

This unusual name was first given to the Schwyll Pool, so named on the 1885 edition of the six-inch Ordnance Survey map, which was formed with several others in that location by an extremely strong flow of water issuing from underground limestone fissures, 'a very copious spring, locally called The Shew Well but usually designated by tourists Ogmore Spring ... the waters, uniting immediately on their emission, at first occupy a space about fifteen yards wide', according to Samuel Lewis in his *Topographical Dictionary* of 1843.

This intriguing form *The Shew Well* bears a superficial resemblance to Schwyll, but it is not possible to ascertain which one of these forms is a variation of the other, should that be the case, because of a complete lack of early documentary evidence. The late Major Francis Jones, in his standard work on the holy wells of Wales, refers to local legends about the *Shew Well*, or the *Shee Well* (an alternative form quoted) and ventures the opinion that it may well be worth noting that the Gower word for a stye is *she*, adding the possibility that the *Shee Well* acquired its name as a well whose waters could cure sore eyes. However, the accretion of legendary material presupposes a reasonably long period of identification which makes the scarcity of early forms of the name even more puzzling.

Be that as it may, in view of the fact that no corresponding Welsh form of the name has hitherto come to light it can be accepted that it is probably English, and because of the lack of dated documentary evidence as to its origin it may be profitable to refer to some comparative evidence.

The basic E *swill*, in its earlier sense of 'a strong flow of water' is worth considering, this being a feature of the emissions from the underground fissures noted above. It is related to the OE verb *swillan*, *swilian* 'to wash out' and river names in the English counties of Kent, the North Riding of Yorkshire and Berkshire which now take the form Swale were earlier, in the sixteenth and seventeenth centuries, *Swyll(e)*, *Swill*. Also Swale Bank in Sussex was 'a place where water ran freely and strongly'. If such a form underlies the present Glamorgan Schwyll, it would appear that it has retained its early modern form, a partial fossilisation, because the local colloquial influence which produced the later form Swale in England would hardly have been present here. The initial consonantal combination *sch-* can be regarded as an orthographical change evidenced in Middle English to denote the sound of *sh* when derived from OE *sc*. To this may be added the tendency for Welsh speakers in Glamorgan to pronounce initial *s-* as *sh-* in certain circumstances.

Southra : Westra : *Soudrey*

The first two of these forms are the names of two farms on the fringe of Dinas Powys Common, the first element in the one as well as in the other being obviously points of the compass to indicate the location south and west, possibly in relation to the Common. They both have a common second element, *-ra* in the modern forms. This is an unaccented colloquial form of ME *rewe*, OE *rēw* or *rāw* 'row', the former suggesting more often 'a row of trees' and the latter, particularly, 'a row of houses' or 'a range of buildings', the modern *row* being similarly derived but with a variant vowel development. Attested forms are: *Southrewe* 1455, *Sowth(e)rewe* 1489 and *Westrewe* 1455, *Le Rewe* 1489, 1492, *Westrew* 1561–2, 1558–1603.

The present writer once believed, in accord with the explanation offered by Dr B.G. Charles for the meaning of Southrow in the parish of Castlemartin, Pembs., that the *row* in each case here was a reference to a row of trees strategically placed as windbreaks, and some tentative oral testimony to this effect was obtained from a very elderly lady. The use of the term *row* in such a context is generally well evidenced but there is another consideration to be borne in mind, for in the area of Dinas Powys Common another *row* is recorded, namely *Northrewe* in 1455. It is not possible to ascertain its location with certainty since the name has not survived and no further documented evidence for it has been found to date.

This brings into consideration the existence of another *Southrew* c. 1600 in the vicinity of the southern gateway of the old borough of Cardiff, a little beyond the southern end of the present St Mary Street, which was John Leland's *Portllongey, in Englische the Ship Gate*, and out of which a street called *Schipmanstrete* (ModE *shipman* 'sailor, mariner') ran in 1321. By the sixteenth century the area had undergone enough urban development, both inside and outside the gate, to be recognised as a suburb of the town. Its name is well-attested as Soudrey: *Soudr(e)y* 1550, *the Sowdrey* 1562, *Sowdrey* 1563, 1570, *Soudrey* 'without the South Gate' and 'within the South Gate' 1666 etc., also as *Soudra* 1723, the gate being *(the) Soudrey gate* 1678–9.

This name is no longer current, but it has been wrongly interpreted as an example of early Scandinavian influence on coastal place nomenclature in south Wales, as if it were compounded of ON *suðr* 'south' + the ON suffix *-ey* 'island', seen in names like Bardsey, Ramsey, Anglesey etc. However, in view of the complete lack of very early recorded forms of the name which might have been expected if it were of Scandinavian provenance, and having regard to the fact that the gate cannot have existed before the foundation of the borough of Cardiff c. 1100, or, indeed, its enclosure by earthworks, palisade and fortified gates, what is more likely is that it evolved as a consequence of the later influence of English West Country and southern county dialects on place-names and minor names in the Cardiff area and the Vale of Glamorgan. Specific reference can be made to the development of the consonantal grouping *-dr-* from *-thr-* seen in the form Drope, from *Throp(e) (Thorp)* in the parish of St Georges (pp. 66–7). The form *Soudra* 1723 would thus correspond exactly to the Dinas Powys *Southra*, the forms in *-ey* having a variant vowel quality in the unaccented final syllable.

It appears, therefore, that ME *rewe* 'row' in this connection is from OE *rāw* 'a row of houses', virtually 'a street of houses'. A deed of 1616 refers to 'ground called the Rowe in Sowdrie' which suggests not only that it was a familiar enough term in the area but also that this is the sense in which it is to be understood in the name *Soudrey*. There were several Middle Rows in Cardiff as well as others such as Bakers' Row, Court Colman Row (itself near the South Gate) etc.

In view of this consideration it may now be profitable to await further investigations into the possible existence of remains of structures to the north, south and east of Dinas Powys Common which would confirm or otherwise that the rows evidenced in the names Southra, Westra and *Northrewe* were of buildings rather than of trees.

The surviving form of the name Coldra, east of Newport, Mon, now a farm name (once the name of three properties, one being the substantial Coldra House) and now also used to name the roundabout at the well-known junction 24 on the M4 motorway, may well have the same suffix, being *Coudrey* 1100–35, *Coudray c.* 1291, *Coudre* 1322, *Cowldre* 1566, *Coldrey* 1596–7 etc. The first element, if it is the OE *cald, ceald* 'cold', suggesting the bleak and exposed location of the original holding on the high ground overlooking the location of present road junction.

Splot : Y Sblot

At least eight examples of Splot as a field or farm name are recorded in the Vale of Glamorgan. It is also found in Gower and Pembrokeshire. Of all those to be found in Glamorgan perhaps the best known example is that of Splot in Cardiff, now the name of a densely populated urban area which constitutes the major portion of the southeast of the city. Still a matter of some disparagement, the name requires a measure of rehabilitation.

The form *splot* and its related form *splat* are variants of the common E synonymous *plot* and *plat*, which are self-explanatory, the initial *s-* being a feature of a number of southern English dialects, yet another example of this particular influence on place-name forms along the south Wales coast, as has been noted elsewhere in this volume.

In the case of the Cardiff Splot, the present urban area was originally flat and largely featureless moorland extending down to the Bristol Channel and the name was first given probably to a portion of this land. It became the name of a holding and persisted as the name of that holding when it became part of the estate of the bishop of Llandaf in the thirteenth century. However, this, most certainly, was not the occasion for its acquiring its name from 'God's plot' as suggested in a recent publication.

By the sixteenth century it had been leased to William Bawdrip (or

Bawdrem to the Welsh-speaking inhabitants) of Odyn's Fee in the parish of Pen-marc, but a native of Somerset. John Leland is able to speak of Splot as *a maner place belonging to Baudrem c.* 1536, this being Rice Lewis's *faire house neere Cardif c.* 1596. In the following century the property was divided into two farms, the upper and the lower, and a portion was sold to Edward Lewis of Y Fan, Caerffili, in 1626. Bawdrip's manor house was the upper farm and we are told that even as recently as the 1860s 'it stood out in the country, all alone except for a barn opposite. The house (Splott House in 1843–50) became the Great Eastern Hotel (in Metal Street) and the barn is now (1905) replaced by the Metal Street School'. The hotel still stands but Lower Splot was further south and east near St Saviour's, Splot Road, the vicarage being reputedly the old farmhouse.

Over the years there has been little consistency in the spelling of the name, documentary forms seemingly being a matter of scribal preference, particularly the form in *-tt*, which is not strictly necessary. Variations such as *Splattye* 1568, *Splote* 1726 etc. also occur, but *Splott* is the predominant form in documentation. From *c.* 1542–3 onwards the name is frequently made more specific by adding the def.art., *the Splott* 1542–3, 1596–1600, 1658 etc., and by the present time a cymricised equivalent, *Y Sblot*, is used in Welsh-medium contexts although it was Thomas Bawdrem *o'r Splott* ('from the Splott') to whom the poet Dafydd Benwyn sang his praises at the beginning of the seventeenth century.

Stormy Down

More than one unfortunate motorist trapped in drifting snow has had occasion to reflect upon the open and windy nature of that stretch of the M4 motorway which rises gradually eastwards from the vicinity of the village of Pyle (Y Pîl). One wonders whether they would have realised that the land which extends from the broad valley of the Ffornwg stream up which the road ascends is called, aptly it would seem, Stormy Down.

On the northern side of the road there once stood the small settlement of Stormy. In the twelfth century, at a point about half a mile due east of the present Stormy farm and a similar distance north of the motorway stood its castle on its motte, the remains of which are still in existence at National Grid reference SS 8458 8153, but the place did not acquire its name from its situation in such an exposed location. The same is true of the open expanse of the surrounding Stormy Down to which it gave its name and which has as its second element the English *down*, OE *dūn* in its basic sense of 'elevated, open ground'.

In fact, the settlement bore the surname of its Anglo-Norman founders which appears as Esturmi, Sturmi, of which Stormy is a variant form, a surname which is well evidenced across the south of England and in Gloucestershire in the eleventh and twelfth centuries. This particular holding was *terra sturmi* in 1138–70, upon which stood *villa Sturmi c.* 1170, later *Sturmieston* 1234 'Sturmi's homestead, or farmstead' with which was associated a chapel or church, *ecclesia de villa Sturmi*, the site of which has not been found.

This unproductive land, nowadays scrub-covered and scarred by limestone workings, which one twelfth-century document described as being 'in a solitary place where no man has ever ploughed before' (*in solitudine in terram quam nemo unquam prius araverat*) was granted to Margam Abbey by Geoffrey Sturmi *c.* 1175 and by 1261 the settlement is specifically referred to as a grange of Margam which by 1347 had been divided into two parts, they, in turn, having both come by 1543 into the hands of Sir Rice Mansel on the Dissolution of the monasteries.

The site of the grange was once thought to be that of grassgrown banks to the immediate south of the castle motte, but the Royal Commission on Ancient and Historical Monuments notes a lack of evidence to support this view and prefers a location further to the south-east at SS 8472 8139 where there are remains of three rectangular buildings in an enclosure for which the architectural and ceramic evidence suggests a medieval date and which may even point to a reuse of the grange by sixteenth-century occupants.

Swansea : Abertawe

There is no linguistic or semantic connection between these alternative names, one being clearly a Welsh name and the other generally recognised as having had a Scandinavian provenance, thus providing supportive evidence for what other sources reveal to have been a period of Viking attacks on the Welsh coastline from across the Irish

Sea, by way of Ireland, from the ninth to the eleventh century.

The Welsh form Abertawe has the elements *aber* 'mouth of a river' here to denote the place where the river Tawe runs into the sea. The old form of the river name was *Tawy* (*Taui, Tauuy* c. 1150, *Tawuy* c. 1200, *Tawy* 1203, 1336 etc.) which may be a by-form of W *Tafwy*, Br **Tamouio* (having the Br root-form **Tam-* like the Cardiff Taf 'water, flow', see p. 25). The place-name is recorded in Welsh sources as *Aper Tyui* c. 1150, *Abertawi, Abertaui, Abertawy* 12c., *Aber Tawy* c. 1300, *Abertawy* 1455–85, *Aber Tawe, Aber Tuwi* 1606 etc.

The form which now appears as Swansea is accepted on firm grounds to have the ON pers.n *Sveinn* as its first element and either of two elements about which there has been a division of opinion, namely ON *ey* 'island' or ON *sær* 'sea, ocean', as the second. Of the two, it may be preferable to accept the former in view of the fact that ON *ey* is a comparatively well-evidenced element in coastal island names of Norse origin in Wales such as Bardsey, Anglesey, Ramsey, Caldy etc. This ties the name down to to a specific topographical feature which is associated with a named person to show possession or overlordship. What exactly, or where the 'island' is or was is a matter which has been vigorously debated. In fairly recent times, some investigators, the present writer included, have been criticised for having accepted too readily the idea that it was an island that may once have existed between the two arms of the river Tawe at its mouth. Be that as it may, an estuarine location must have had features that could have 'fitted the bill', as the saying goes, always keeping in mind the fact that OE *ēg* 'island' was often used in settlement names for areas of raised ground in wet surroundings and on river banks subject to inundation to form watermeadows, like the W *ynys* (see p. 224).

Of greater importance is the realisation that of the two forms it was not the Welsh form which was adopted by the Normans to name their settlement around the castle, the existence of which is recorded in 1116. A hoard of coins found near Rhiwbina to the north of Cardiff in 1980 contained a number which were minted at Swansea *c*. 1140 and bore abbreviated forms of the name: SWENSI, SWENS, SVEN, SWENI, SVENSHI. William de Beaumont, the earl of Warwick and grandson of the first Norman lord of Gower, bestowed the first charter of rights and obligations upon the burgesses of *Sweynesse* 1153–84 (copy in an early fourteenth-century compilation), this being the first documented form of the name which subsequently appears as *Sueinesia* 1187, *Suineshæ*, *Sueinesea* 1190, *Sweineshe c.* 1191, *Suenesel* 1193, *Sweineshea* 1210, *Sweynesia c.* 1214, *Swenese* 1235 etc., until the forms *Swanzey* 1598–1600, *Swansey* 1331 and *Swansea* 1530 make their appearance.

The question which defies a clear answer, however, is to what kind of Scandinavian settlement was the Norman borough a successor? Place-name evidence along the Welsh coast implies some limited measure of settlement, perhaps, such as trading posts and the like but much of it suggests merely what might have been expected in the way of contact with renowned seafarers such as the naming of prominent landmarks, headlands, promontories, rocks and islands. Even when a name is confidently accepted as being a compound of ON elements there is sometimes a difficulty in determining the exact definition of those elements because they cannot be evidenced other than in documented sources of a much later date. In such cases conclusions have to be reached on comparative linguistic grounds. Furthermore, in the case of Swansea archaeological evidence of Scandinavian settlement is virtually non-existent, the most recent find by means of a metal detector on the foreshore in Swansea Bay being a ringed pin which may have had a Hiberno-Viking origin. Much the same is true of the ON form *Hundemanby* now seen in the name of Womanby Street in Cardiff (p. 99) where archaeological excavation has revealed no trace of Scandinavian occupation.

Neither is there any evidence to prove which of the two names is the earlier. Perhaps there is a popular tendency to give pride of place to the Scandinavian form notwithstanding the fact that whatever may have been the nature of the feature or establishment so named it was located, together with the Norman borough which came in its wake, in the southern portion of the Welsh commote of *Gwhyr*, later *Gŵyr* (anglicised *Gower*). This was originally part of the Welsh kingdom of Deheubarth known as Ystrad Tywi and one of the three cantrefi of the area, probably known as the cantref of Eginog. The extent of later anglicisation, including its effect on place nomenclature which saw *Llanilltud Gŵyr* becoming Ilston and *Llandeilo Ferwallt* yielding ground to Bishopston, has obscured its former Welsh identity from modern eyes. The near-contemporary writings of Giraldus Cambrensis, particularly the account of his journey in Wales with Archbishop Baldwin in 1188 and his *Descriptio Kambriae* of 1194, are of no great help in that they do little to differentiate between the castle and the town in the matter of their nomenclature. In the former he refers to the castle of *Sweineshe*, *Sweinesie*, which is called *Abertawe* in Welsh (*Abertau* in one version) which he explains as *casus Tawe fluvii*, 'the fall of the river Tawe' (into the sea), and in the latter he reverses the statement, as it were, by referring to the castle of *Abertawe*, called (he says) in English, *Sweynesia*, this, of course, well after the foundation of the borough and probably in order to explain that both names refer to the same place. A similar ambiguity is seen in the Welsh medieval chronicles. In the portion of the *Annales Cambriae* compiled about the end of the thirteenth century Rhys ap Gruffudd's attack on *Abertawi* (the town) in 1192 is recorded and that Llywelyn ap Gruffudd took the castle of *Abertaui* in 1216, but a later insertion notes that in the year 1287 Rhys ap Maredudd burnt the town in the words *combussit villam de Sweynese*.

Finally, in the Welsh versions of the *Brut y Tywysogion*, the oldest of which is no earlier than the fourteenth century although based on a lost original Latin chronicle compiled towards the end of the thirteenth, the form *Swansea* does not appear. The castle is referred to as *y kastell a oed ... yn abertawy*, 'the castle which was in Abertawe', *kastell abertawy*, and *castell a oed ossodedic yn ymyl Abertawy*, 'the castle which was raised near Abertawe'. Further, an intriguing dimension is added to the point at issue with the appearance in these sources of a third name, for in every reference to the castle in the *Brut* other than those noted above it is called *castell Seinhenyd*, or *castell Sain Henydd*, and on one occasion, *castell Abertawy a Sainhenydd*, 'the castle of Abertawe and Sainhenydd'. In another section there is a reference to the castle of Oystermouth (Ystumllwynarth) on the coast a little to the west of Swansea, and the entry has its name as *Ystum Llwyniarth yn Sein Henydd*, 'Ystumllwynarth in Sainhenydd', as if Sainhenydd were a district in which the castle was located. However, in yet another entry Rhys Ieuanc is said to have 'made for Seinhenyd' (in Professor Thomas Jones's translation) in a context in which it can only mean the garrison town of Swansea. To equate this form with that of the probable name of the cantref of Senghennydd north of Cardiff is not possible on philological grounds alone, despite the fact that one version of the *Brut* refers to Senghennydd as *Sainhenyd*.

Clearly, a difficult problem of identification is indicated here compounded, it would seem, by errors of transcription or translation or both, and on which the location of a dedication to the ubiquitous saint Cenydd in Gower, not too far removed from Swansea, may well have had a bearing. This is a problem to which a solution has not yet been found.

'Sweyne's Howes'

This is the name which appears on Ordnance Survey maps to identify the location of two megalithic burial chambers now in cairns of small boulders, much disturbed, on the eastern side of Rhosili Down in Gower. It is also used as the caption to the Royal Commission on Ancient and Historical Monuments' entry on the monuments in its Glamorgan Inventory.

On the 1883 Ordnance Survey six-inch map the name appears as *Sweyne's Houses*, and the historian of west Gower, the Revd J.D. Davies, writing in 1877, maintains that this is 'in reality Sweyn How' and suggests a Scandinavian origin for the name. His note on the subject, however, is entitled 'The Swine House', but he does not enlarge on the possibility that this form of the name was the one in common usage at the time of writing. That this was so is the opinion favoured here.

It does not seem to be a name of long standing. Conventional sources provide no early evidence of it despite the antiquity of the monuments so named. The ruined chamber of the north cairn still has two supporters with the capstone which lay on top having slid off to lean against them at its western end. With the capstone in its original position, like other monuments of this category, it is not difficult to imagine a popular interpretation of the remains as some kind of rough shelter or den. It is a known fact that *house* is popularly used in relation to barrows in English place-names. In Wales a similar popular attribution is made in relation to animals which might have frequented such locations. The remains of the chambered tomb at Maesyfelin, St Lythans was named Gwâl y Filiast, W *gwâl* 'den, covert, lair' + *miliast, milast* 'greyhound bitch', this name being also in use in the parish of Llanilltud Faerdre. Llech y Filiast, with W *llech* 'stone slab', Llety'r Filiast, with W *llety* 'lodging, temporary dwelling', and Carnedd y Filiast, with W *carnedd* 'cairn, mound' are others of a similar kind evidenced in Wales.

Among other animals featured in such names is W *hwch* 'sow', particularly in the compound Twlc-yr-hwch, with W *twlc* 'hut, cabin, shelter' borrowed from Ir *tolc* (the cognate W word being *twlch*, pl. *tylchau*). About three miles in a direct line from the Swine House a slope on Ryer's Down, in the parish of Cheriton, subsequently the name of two farms, was called Rhiw'r-hwch with W *rhiw* 'hill, ascent, slope' evidenced as *Reuroch* 1282, 1314, *Rewrough* 1315, *Rewroch* 1323, *Rewrozth* 1516, *Rywrhwgh* 1516, 1588, *Rhyer hwch*, *Rhyier-hwch* 1598–1602, the land being rough and open common similar to Rhosili Down, on which swine were undoubtedly wont to forage. Indeed, the name *Boarspitt* 1650 in the not-too-distant parish of Oystermouth is also reminiscent of the same species.

It would certainly seem that in such circumstances the Swine House could be regarded as a natural popular appellation for a hut-like stone structure frequented by swine. As for ascribing to it a Scandinavian provenance, one suspects a contrived antiquarian influence in the creation of the form Sweyne's Howes, possibly based on the analogy of the pers.n. *Sveinn* evidenced in the name Swansea (p. 182) + ON *haugr* 'burial-mound, tumulus, hill' which gives the E form *howe* in later names, but here given a pl. form to cover both surviving structures. Added to this may have been the belief that two other names recorded in Glamorgan contain the same pers.n., namely the *gardinum Sweini* of *c.* 1200 in the fee of Newcastle, Bridgend, and, far less likely, Wainsill, Cadoxton, Barry, which is not recorded, as *Sweins Hill*, until the very late date of 1762, as far as is known, but which appears thereafter as *Swinshill* 1784, *Swinehill* 1785, *Swineshill* 1789, *Waunshill* 1896.

Talycynllwyn

This name, that of a farmhouse on the banks of the Camffrwd which runs into the Llwchwr river north of Pontarddulais, has been subjected to several attempts at interpretation based mainly on the assumption that the second element *cynllwyn* is the W masculine noun which has the sense of 'plot, intrigue, treachery' or 'ambush', thus giving inventive minds scope for vivid explanations. In any case, if the first element is W *tal* 'end, forward end' as in *talcen* 'forehead, brow' or names like Tal-y-bont, Tal-y-llyn, Tal-y-sarn etc., as is very likely, it is difficult to see how such an element could occur in a place-name in combination with an abstract noun like *cynllwyn* unless it should happen to have been descriptive of an area, possibly wooded, suitable for, or noted for having been the scene of an ambush.

Early forms of the name are not numerous but those that are evidenced point to a different interpretation: *Taly kelyn llwyn* 1584–5, *Talyklynllwyn* 1613–14, *Tal y Clyn llwyn* 1692, *Talyclynllwyn* 1765, all clearly combining the elements *tal* + the W def.art. *y* + W *celyn* 'holly' + W *llwyn* ' grove, copse; bush' giving the sense 'the end (head) of the holly grove'. The order of the last two elements is the reverse of that which is more often seen in the common Welsh place-name Llwyncelyn where the accent is firmly on the penult. In the form *celýnllwyn* it is the second syllable of *celyn* which becomes the accented penult which results in the loss of the unaccented preceding vowel in common parlance > *c(e)lýnllwyn* and *clynllwyn*. This can be compared with that which occurred in the case of the place-names Clynnog Fawr and Clynnog Fechan, Caerns., Anglesey, < *celyn* + the ending *–og* which signifies a place where there is an abundance of the named tree or plant (p. 72) and in the lenited form after a lost W def.art. Glynnog (Y Glynnog) in the parishes of Llantrisant, Llanwynno and Pen-tyrch. Similarly, the name Clenennau (Clenna in local speech), Caerns., a plural form of *celynen*, namely *celynennau*, which signifies 'a number of individual holly trees' in the opinion of Sir Ifor Williams.

However, in the form Talycynllwyn it is clear that a further loss of the consonant *-l-* has occurred in the first syllable of the combination *-clynllwyn* which is more difficult to account for unless it was a deliberate attempt to introduce a current word, *cynllwyn*, in order to give the name a semblance of meaning, for *cynllwyn* does occur as an element in other names. Near the old colliery site at Penallta in the parish of Gelli-gaer stands the farm of Cynllwynau, a plural form in *-au* which is Cefnllwynau on some maps and this is confirmed by the forms *Keven lloyne* 1566, 1576 and *Keven y lloyne* 1584, that is W *cefn* 'back, ridge' + *llwyn* which has assumed its plural form by the present time. It is known that *cefn* as a first element has developed the forms *cen-* and *cyn-* in the vernacular, as in the forms Cencoed and Cyncoed (p. 83) so that *Cynllwyn(au)* < *Cefnllwynau* is entirely possible, but there is no trace of *cefn* evidenced in the collected forms of Talycynllwyn.

Other Glamorgan names which have *cynllwyn* as an element are Llwyncynllwyn and Cynllwyn-du in the Rhondda valley. The former is not likely to be a tautologous form with *llwyn* repeated as first and last elements, whereas to regard Cynllwyn-du as a possible derivative of *cefn-llwyn-du* would be reasonable were there the evidence to corroborate this. This is lacking, and so is any evidence concerning the identity of Cynllwyn (*Cinluin* in Old Welsh orthography) which is referred to in the *Liber Landavensis* as being in the bounds of Llanwytherin (Llanvetherine on the map), Mon. Meanwhile, there is no supporting evidence for Coed y cynllwyn (with W *coed* 'wood, trees') which Iolo Morganwg claims in one of his manuscripts to be the Welsh name of Wrinstone in the parish of Wenvoe, and it is unlikely that Coed Cynllan in the parishes of Llanharan and Llanhari is in any way connected. Such problems evidently require further consideration.

Tirergyd

On the slopes of Mynydd Aberdâr (Aberdare Mountain) in the area formerly included within the Forest of Llwydcoed and not far from the site of the old Aberdare Ironworks stood the farmstead of

Tirergyd. Indeed, included in an endorsement of the lease which effectively conveyed ownership of the Llwydcoed Forest to two English industrialists in 1787 is a note to the effect that 'there was a farm called *Tir Ergid* immediately above a portion of the Llwydcoed estate of which it was essential to lease the minerals if the iron-works were placed on the spot ... '. This also confirms that the farm was located on high ground. The earlier forms of the name available are *Tyr yr Ergyd* 1632, *Tyir yr Rergyd* 1633, *Tyre yr Ergyd* 1638, *Tire yr Ergid* 1651, 1688, 1715, *Tir r ergid* 1791 etc. On the first edition one-inch Ordnance Survey map it appears as *Tiererged*, subsequently to assume the erroneous form *Tir-yr-argae* on the 1884 six-inch map.

The tendency has been to see here the W epicene noun *ergyd* 'blow, knock; throw, cast; a distance to which an object (or weapon) will carry' as the second element preceded by W *tir* 'land' and possibly having a veiled reference to some fabled event concerning the casting of a stone or weapon a vast distance by an unknown hero. This is not implausible, of course, but it is instructive to note the use made of this element in other examples of its occurrence, particularly *Ergyd Uchaf* and *Ergyd Isaf* which are the names of two spurs or ridges on Mynydd Margam in the first case and Mynydd Brombil, in the second, both east of Port Talbot and both being the locations of Bronze Age burial cairns at National Grid references SS 8061 8888 and SS 7943 8865 respectively. The locations are spurs which 'project' forward, as it were, from the body of the hills in each case so it is this aspect of the basic sense of *ergyd*, suggesting 'distance, throw' which is retained and adapted to describe a natural feature.

GPC derives the word from *er-*, *ar-*, intensive prefixes + the element *cyd* which occurs in words like W *cyngyd* 'adjoining, contiguous' as an adjective and 'common boundary, conjunction' as a noun, and *encyd* 'moment, short distance of time', notwithstanding that it was once thought to be a form of W *ergyr* 'blow, throw' (having the Celtic root-form **kor-* 'place, throw, cast') by dissimilation, which occurs in Cwmergyr, not far from Ponterwyd, Cards.

Be that as it may, further examples of *ergyd* occur in names which appear to corroborate its use as a term for a topographical feature such as a spur or projection. A clear projection of land at the mouth of the Teifi, opposite Poppit Sands near Gwbert is called Pen-yr-ergyd, and in Llanrhaeadr-ym-Mochnant, Denb., there is Ergyd-y-gwynt (W *gwynt* 'wind') suggesting an exposed situation. It also occurs with personal names, probably to indicate possession, as in Ergyd Non, Llansanffraid, Cards., and Ergyd Ronw, which is *Ergid Gronow* 1595, *Tyddyn ergid Gronowe* 1592–3, 1623, *Dol y ronw alias Ergid Ronw* 1691, with the personal name Gronw < Goronwy, near Dolgellau, Merioneth.

Tir-shet

Near Rhyd-y-fro in the civil parish of Rhyndwyglydach stands a farm of this name which has been the subject of many enquiries as to its meaning. As far as is known, there seems to have been no incident worthy of note in its past yet the name is indicative of a change which befell the status of the land after which it has been called.

Exactly when such a change occurred is not known but it certainly had occurred before 1613 because its name was recorded as *Tyr Cheate* in a will of that year, subsequently *Tir Ziet* 1664, *Tiryet* 1722 (possibly in error for *Tirsyet*), *Tur Shet* 1783, *Tyr y sliet* 1807 (probably *siet*), *Ty'r shet* 1846.

The earliest form implies strongly that the second element, after W *tir* 'land', is a W borrowing of the English term *escheat*, ME *(es)chete*, of OFr, earlier Latin, origin. The unaccented first syllable was lost at a fairly early stage (cf. the borrowed form *stent* for E *extent* 'valuation of an estate' and *sart* for *assart* (p. 165)), and the Welsh form retained the long vowel quality of the syllable in the form *siêt*, also *siêd*, until fairly recent times when it seems to have been shortened in the vernacular, as suggested by the *shet* of 1783. It may be that it is the form with the long vowel quality retained that survives in the name of the southern portion of the medieval road from Margam to Cynffig (Kenfig) known as Water Street for some distance and then *Heol y Sheet*, so named in 1756–7 and having some remote connection with land held in the area which cannot now be specified.

Tir siet is, therefore, 'escheated land', that is, land which has reverted to a lord when a tenant was either guilty of a felony or when he died without heirs. There is evidence that in some parts of Wales in the Middle Ages such land reverted to the possession of the local community. However, in common speech, *tir siêd* was a term increasingly used to signify land which had no owner and consequently acquired the wider related adjectival sense of 'empty, worthless, desolate' etc. This is the meaning which the word has in the lament of the poet Iolo Goch in the fourteenth century on the condition of the isle of Anglesey after the death of the sons of Tudur ap Gronwy, *Mon aeth, ysywaeth, yn sied* ('Mon, alas, became desolate').

In Welsh, the word also developed the meaning of 'forfeit', but here it is an example of a technical landholding term being incorporated as an element in a place-name much as was the case with *assart* in Cwrt Sart near Neath.

Tontrycwal

Around the embanked hill of Caerau, Llantrisant, there is a cluster of farms which still resist the spread of urbanisation. One of these is Tontrycwal which stands by the side of the narrow road from Cross Inn to Rhiwsaeson and on the banks of a little stream which runs into the river Clun.

The name is not printed on the majority of modern maps but on the first edition one-inch Ordnance Survey map in 1833 it appears as *Tondrygwal* and the form of the name on the entrance gate of the present house is *Tondrugwaer*. The form which appears at the head of this note is that which was heard on the lips of the older inhabitants of the district.

Again, as far as is known, the place does not seem to have had a startling history, but its name is unusual and it did have its niche in the history of the local woollen industry for there was a small grist mill associated with the farm which was later adapted for use as part of a small woollen factory, not that this activity is in any way reflected in the name.

The name is reasonably well documented and from the available evidence it is clear that it was not until the first half of the eighteenth century that the final *-l* makes its appearance: *Tondrigwel* 1728, *Ton Drwgwall* 1771–81, *Ton-drugwall* 1874 etc. Earlier, including the earliest form recorded, the final letter was clearly *–r*: *tonne y drewge gwayr* 1570, *Ton y d(r)wg Gwaer* 1588, *Tyr tonn y drwg wayr* 1630, and in a Welsh rent-roll among the Bute estate papers of *c.* 1625 is the entry *syddyn y elwyd tonn y drwgwayr* 'a holding called *tonn y drwgwayr*'.

A rendering of the original form of the name in modern orthography is *Tonydrygwair*, with the loss of the W def.art., which constitutes the second syllable, in common speech > *Tondrygwair*. This combines the elements W *ton*, a common place-name element in south Wales which has the basic meaning of 'skin, crust, surface' applied topographically to indicate land that has not been 'broken' or ploughed, 'lay-land, grassland' + the W adj. *drwg* 'bad, poor' (assuming the form *dryg-* by vowel mutation when a syllable is added) + W n.m. *gwair* 'grass, hay' (represented in documents by non-Welsh scribes as *gwayr-*, *gwaer-* etc.), the whole indicating a portion of uncultivated land which was poor grazing land or produced a poor crop of hay.

The colloquial form *Tontrycwal* shows a number of features which characterised the Welsh dialect of the area. Provection of *-d-* to *-t-* at the beginning of the second syllable and of *-g-* to *-c-* at the end of it and the reduction of the diphthong *-ai-* to *-a-* in the final syllable of a multisyllable combination. Noteworthy too is the substitution of final *–l* for *-r* which is reminiscent of that which occurs in some Welsh borrowings from English, such as *dresel, dresal* < E *dresser*, *rasal* < E *razor*, and *cornel* < E *corner*.

Tredelerch

This is the Welsh name of the old demesne manor of Rhymni (anglicised Rumney, Rhymney etc.) on the eastern outskirts of Cardiff, once part of the lordship of Gwynllŵg, in Gwent. Its castle is now recognised as having stood above a bend in the Rhymni river and on ground which rises from, and overlooks the level expanse of the estuary of that river to the south of Rhymni Bridge. Both the castle and the bridge of *Remny* are recorded in 1184 but evidence from excavations suggests the possibility of its foundation soon after the Norman occupation of Glamorgan and Gwynllŵg c. 1093.

The manor was virtually coterminous with the parish of Rhymni, its church being named in the ecclesiastical tax returns of 1254 and c. 1291 as *ecclesia de Renny* and *ecclesia de Rempney* respectively, but having no specific name of the type prefixed by an element such as W *llan* or *eglwys* + the name of a saint. The dedication is to St Augustine, the church having been given to St Augustine's Abbey, Bristol, by the second earl of Gloucester (1147–83), who was also lord of Glamorgan (although there exists a record of a church of St

John of Rumney, of which nothing is otherwise known, before that time).

However, just under a mile east-north-east of the church stood the 'great house' of the manor, Llanrumney Hall, the home of a junior branch of the Morgan family of Tredegar House near Newport from the seventeenth century: *Lanrumpney, Llanrhymney* 1624, 1626, *Lan Rumney* 1650, *Lanrompney* 1651, *Lanrumney* 1653, 1677, 1708, 1716 etc., but this name has no religious significance. The prefixed *llan-* here is not the common W *llan* 'enclosure, cell, church' but an erroneous restoration of the lenited form of an original W *glan* 'bank, shore' + the river-name Rhymni in reference to a location by the riverbank. It also serves as a dire warning to those who would attempt to interpret *llan-* names in Wales uncritically. We can postulate an original *Glanrhymni which became *Lanrhymni by lenition of the initial consonant after those Welsh prepositions most frequently used with place-names in daily converse, *i* 'to' and *o* 'from', and after a period of usage the lenited form sometimes gained currency as, no doubt, in this case, only to be wrongly interpreted as an error for a form in *llan* which, of course, would also have been lenited, and restored to what was assumed to be the correct form *Llanrhymni* (Llanrumney etc.).

This particular substitution of elements occurs with a frequency in Wales which can be called common. Two examples which occur at not too great a distance from Rhymni are Llancaeach and Llanbradach. Both have their origins in the names of locations on the banks of small streams. The Caeach (with *-ae-* > *-ai-* in local speech) is the stream used as a line of demarcation between the two commotes of Senghennydd Uwch Caeach and Senghennydd Is Caeach and also occurs in the name of the restored house of Llancaeach Fawr raised near the banks of the stream in the sixteenth century and recorded as *Glancayach* 1619, 1645, 1649, the lenited form appearing after the preposition *o* 'from' in Welsh texts, *o lann Kaeach* 1550–1600, *o Lan Kaeach* 17c. etc.

Llanbradach became the name of the village connected with the colliery which was worked from 1894 to 1961 but earlier it was the name of the home of the Thomas family, Llanbradach Fawr. This originated as a holding named Blaenbradach (Rice Merrick has *Blayn Bradach* 1578) near the source or upper reaches (W *blaen*) of a small stream called Bradach on the slopes of Mynydd Eglwysilan, the name of the stream being, possibly, a personal name of Goidelic origin following the naming pattern of a number of streams and rivers in Wales.

The major portion of the parish of Rhymni was brought within the boundary of Cardiff in 1938 followed by Llanrhymni (Llanrumney) in 1951, the bulk of the housing which forms the suburb of Llanrhymni having been built on the parkland of Llanrhymni Hall.

It is not possible to ascertain whether or not the W form Tredelerch was the name of the original settlement which may have been the core of the manorial unit of Rhymni but there is no doubt that it is a name compounded of W *tref* 'holding, farm, dwelling' + the name of its founder or original owner, *Telerch*: *Tredelerch* 1536–9, *Tref Delerch* 1606, *Tirdelarch* 1698, *Tredelogh* 1698, *Tredeler* 1857. W *tref, tre*, is ultimately derived from IE **treb-* 'dwelling, habitation', Bret *trev*, MCo *tre*, OIr *treb*. Although it has developed a similar meaning to E *town* it originally meant 'dwelling, homestead, farmstead' and appears in the compounds *cartref* 'home', *pentref* 'hamlet, village', *hendref* 'old homestead'. It is in the sense of 'homestead, farmstead' that it occurs in most place-names which are not of modern provenance, as in the case of Tredelerch, but the second element of the name has been misinterpreted with consequences that are still evident.

The very old personal-name *Telerch* is composed of two elements, the W honorific or hypocoristic prefix *ty-* (cf. *Ty* + *gwrog* > *Ty wrog* > *Twrog* in Llandwrog, or *Ty* + *eliud*, or the pet-form *eilo*, MW *eliaw*, *Ty-eliaw* > *Teliaw* > ModW *Teilo* in Llandeilo) and an old Welsh personal-name *Elerch*, earlier *Eleirch*, of uncertain origin, but which can be found as a simplex place-name in Cards., and is named in a poem by the medieval poet Dafydd ap Gwilym in which he refers to his journeying *ar draws Eleirch*, 'across Eleirch' (using the earlier form) in his own habitat. Consequently, because of the known fact that Dafydd ap Gwilym had a patron, Ifor ap Llywelyn (or Ifor Hael 'Ifor the generous') of Gwernyclepa in Gwynllŵg, barely three miles east of Tredelerch's location on the Rhymni, the assumption was made that the second element in Tredelerch was *Elerch* (not *Telerch*) or, more particularly, *Eleirch*, the earlier form, and that this was an alternative name for the river Rhymni. It was but a short step to assume wrongly that the form *-elerch*, *-eleirch* was related to the plural form of W *alarch* 'swan', namely *elyrch* for which, as it happens, there is on record a variant form *eleirch*.

It was from such a misunderstanding that the widespread belief arose that the name of the river Rhymni is to be interpreted as 'the river of (the) swans'. Indeed, William Owen (Pughe) in his Welsh dictionary of 1803 under a strange entry *Tred-*, gives *Tred Eleirch* as the name of 'a district on Rumney in Monmouthshire, called by the English Swanton'. No documented record of *Swanton* is known to the present writer, nevertheless the interpretation of Rhymni as 'the river of (the) swans' persists in some quarters.

Tredodridge

The small village or hamlet which bears this name stands between Hensol and Pendeulwyn (Pendoylan) in the Vale of Glamorgan where the names of places beginning with W *tref* in its early sense of 'farmstead, homestead' occur quite frequently, as in Trerhingyll, Trecastell, Tre-groes, Trehwbwb (p. 199) etc. Some are original Welsh names but many are Welsh forms of Anglo-Norman or English names ending in *–ton*, OE *tūn* 'enclosure, farmstead', later 'village, estate', like *Tresimwn* for Bonvilston (Simon de Bonville was the Anglo-Norman lord) or *Trefflemin* for Flemingston, *Tregawntlo* for Candleston (p. 197).

It would not be unreasonable, therefore, particularly in view of the un-Welsh appearance of the second element of Tredodridge and the common occurrence of personal names or surnames as the second element of names beginning with *tref-*, to assume that this name follows the same pattern. It is also relevant to recall that the Vale of Glamorgan received its fair share of immigrants from over the Bristol Channel over the years and that *Dodridge* is the name of a place near Crediton in Devon and could also be, therefore, a locational surname.

Some names and words of English origin which end in the combination *-idge* are borrowed by Welsh scribes and the ending represented in the Welsh forms by *-is*. *Cwbris* is recorded for Cowbridge, *petris* for partridge, and in the Gwentian dialect the *-is* combination also becomes *-ish* (as in *prish* < *pris*, *mish* < *mis* etc.). This may well be the explanation for the earliest form available for Tredodridge, namely *Tredodrish* 1742, 1797, with provection of the medial *-d-* in *Tredotris(h)* 1799, 1813, another dialectal characteristic, the 1833 Ordnance Survey one-inch map having, surprisingly, *Tre-dotrus*.

The evidence for regarding the second element *-dodridge* as an

English surname is thus reasonably strong but it is also clear that Tredodridge was not the original form of the name. In 1573 *Redodrys* is recorded and an inexact form *Rydd Addris* in 1649. Thereafter, in tandem with *Tredodris(h)* as it were, 1742, 1797, 1799, 1819, *Rhydodris(h)* appears consistently in the eighteenth century, 1737, 1747, 1767, and persists until 1841 in the form *Rhydodridge*. It would appear, therefore, that *Rhydodris* is the original form, with W *rhyd* 'ford' as first element, and it is significant that a brook, Nant Tredodridge on the 1884 OS six-inch map, runs through the hamlet and under a small bridge to which the ford may well have been the precursor, 'Dodridge's ford', for as far as is known, no record of a Welsh pers.n. *Odris* exists, and no indication that it could be a variant form of the well-known *Idris*.

Tref-y-rhyg

Two well-known residences near Tonyrefail bear this name but on current Ordnance Survey maps the forms are Treferig House and Treferig Isha (*isha* = *isaf* 'lower') whilst the first edition one-inch map of 1833 has *Trefereeg*. On the other hand the oldest recorded forms available occur in the oft-quoted survey of the earl of Pembroke's lands in Glamorgan, 1570, and they are *Tre y ryg ycha* (*ycha* = *uchaf* 'higher') and *Tre y ryg Issa*.

Tref-y-rhyg Isaf was probably the house of John Bevan in the second half of the seventeenth century and early eighteenth. He adopted the beliefs of the Quakers in 1688 and gave land to raise a meeting-house in the vicinity in 1692. He was the main pillar of the establishment here and at Llantrisant after spending time in Pennsylvania between 1682 and 1704.

The name is well evidenced and most of the forms which occur in documents which can be relied upon support the view that its two main elements are W *tref* 'homestead, farmstead' + W *rhyg* 'rye', possibly borrowed from the OE *ryge* 'rye', that is 'farmstead where rye is grown'. In addition to the 1570 form, there occur *Trefyrhug*, *Trefyrhyg* 1738–46, 1799, *Treferig ucha* 1738, *Treverhig* 1746, *Trevoryg* 1756–7, *Tref-y-Rhyg* 1768 etc. Similarly, names like Croffty-yr-haidd and Gelli'rhaidd with W *haidd* 'barley' occur in the same area. If we correlate W *tref* with OE *tūn*, *Tref-y-rhyg* would correspond to English names like Ryton (found in Durham, Shropshire and

Warwickshire), Ruyton (Shropshire) and Royton (Lancs.).

There is much variation in the way the name has been presented and treated in the past, the most frequently seen being forms like *Treferig, Treverig, Treverick* which imply that the second element is the initially lenited W personal name Meurig, particularly in its anglicised form Merrick, as in *Trevericke* 1671, *Trefeirig* 1688, and *Treveurig (Mill)* 1844, but the reason for rejecting that possibility is that in local parlance and traditionally the main accent falls on the last syllable, not the penult, as forms written as *Trevorŷg* 1756–7, *Tref-y-Rhŷg* 1768, *Trevereeg* 1783, *Treverhig* 1862, *Treferîg* 1894, though erroneous in some respects, tend to corroborate. It is precisely this accentuation which has led to the unaccented medially situated W def.art. *y* becoming indistinct and influenced, by the secondarily stressed *–e-* of the first element *tref*, to produce what could, and did sound like *-ferig*, assumed to be from Me(u)rig, Merrick.

Tregawntlo : Candleston

This was the name of a manor on the edge of the sandhills in the parish of Merthyr Mawr which has now disappeared beneath the encroaching sand, in all probability, but whose name still survives in that of Candleston Castle, the ruin of a fifteenth-century manor house which stands on an elevated position overlooking the dunes and was tenanted until *c.* 1900. Its Welsh name, Tregawntlo, should prove a sufficient deterrent to those who would still wish to associate the place with candles.

It is, of course, one of the many places in the Vale of Glamorgan which have a name beginning with *tref* in its Welsh form and ending in *-ton*, OE *tūn*, in its English form, both essentially meaning 'farmstead, homestead' in the first instance, later to develop expanded meanings according to how such places developed, indeed, in some cases ending with 'town'.

In the majority of such names the second element is either the personal name or surname of the founder of the settlement, usually of Anglo-Norman origin, such as are found liberally scattered in surviving documentation from the twelfth to the fourteenth centuries. In this case it is the family of de Cantelupo which assumes forms like Cantilo, Cantelo(w), Cantolo etc. and rendered in the Welsh forms of the name as Cantelo, Cantlow, Cawntlo etc. However, what is particularly noticeable is that the direct evidence for the existence of this homestead is not as early as it might have been expected to be, for a papal bull of 1128 confirming the possessions of the church of Llandaf lists among them twelve acres of land, previously taken by force, now restored by William de Cantolo, and other members of the family are conspicuous as witnesses to documents in medieval times.

Both the English and Welsh forms of the name available hitherto are no earlier in their provision than the middle of the sixteenth century: *Cantlowstoune* 1545, *Cantle ston(e)* 1578, 1695, *Cantlostown* 1596, *Cauntleton* 1598, *Cantelowstowne* 1596–1600, *Candleston* 1716–17 etc. and *Tre gawntylio* c. 1550, *Tregantelow* 1559, *Tre gawntlo* 1573, *Tregantelo* 1578, *Tregantlow* 1630, *Treganthloe* 1839. In fact, no direct documentary evidence exists to prove a connection between the family of de Cantelupo and the place so that the place-name evidence here supplies what the historian's conventional sources fail to do.

The contention made by one respected local historian that the de Cantelupo family derived their name from that of the place is dubious for it is reasonably certain that it is of Norman-French origin, a locative surname from one of two places in Normandy, or both, Canteleu (Seine-Inferieure) and Canteloup (Calvados).

With this name can be compared the naming of Nurston near Rhoose in the parish of Penmark which probably combines the surname of the family of Norreys, Norreis (Norris) with *-ton*, this being the Anglo-French *noreis* 'northener', and is common also in the English midlands and the south. Members of the family of le Norreis appear among the *advenae* to the Vale of Glamorgan between the twelfth and the fourteenth centuries and held Pen-llin (not Penllyn) north of Cowbridge in 1262, together with land in Bonvilston and Llancarfan, but no record is otherwise available of their tenure of any property as far south as Nurston. The local Welsh pronunciation of the name was *Nyrstwn*.

Trehwbwb

A favourite tendency among those who attempt to interpret place-names uncritically is to ascribe to them an antiquity which cannot be validated. This has been the fate of Trehwbwb, the name of a modest holding near the village of Llwyneliddon (St Lythan's) in the Vale of

Glamorgan in the eighteenth century but a substantial residence by today.

As far as the form is concerned, it is *Tir Wbwb* in 1762, with W *tir* 'land' as first element, and then *Treoupup Land* 1764, *Trehubub Land* 1785 etc. with W *tre(f)* 'homestead, farmstead' as first element and E *land* (= W *tir*) appended, but the persistent use of that appended element in eighteenth-century documentation, possibly as an 'explanatory' element, and the lack of earlier forms, suggests a name of comparatively recent origin which had W *tir* as its first element. The substitution of one for the other in the case of *tir* and *tre(f)* is common, particularly when the homestead raised upon a piece of land came to be known by the name of that land, in addition to the superficial similarity of the two forms.

The second element *hwbwb* is a borrowing from E *hubbub* 'confusion, disturbance, turmoil' used here and elsewhere to designate land about which there had been some dispute. The evidence for this kind of name is considerable in England, the best example being Debateable Land in Cumbria, earlier *Threpelands*, having OE *þrēap* 'dispute, quarrel, contention' as its first element as in names like Threapwood, Threapwaite etc. In Wiltshire, field-names like No Mans Land, Whose Land, are on record and so is an early *Disputforlang c.* 1200–10 in Oxfordshire. In Wales, the n.f. *dadl* 'argument, dispute' (rather than the earlier sense of 'lawsuit, plea' or 'assembly' and possibly 'place of assembly') or the plural form *dadlau* (also used as a verbal substantive) conveys a similar sense, as in Llwynydadlau (*Lloyn y Daddle* 1594, *Llwyn y Dadley* 17c. etc.) or Tir y Dadlau (*Tire y Dadley* 1638, *Tyr y Dadley* 1671, 1778 etc.) in the parish of Aberdâr, or *Tir Llwyn y Dadley* 1660, Pembs. Again in the parish of Llwyneliddon a *Cae Scandal* is recorded in 1660, and on the northern slopes of Caerffili Mountain stands the dwelling house of Cwmwbwb, which was *Cwm Ywbbwb* 1799.

This later form *Ywbbwb*, or *-iwbwb* in standard Welsh orthography, is a colloquial variant of *hwbwb*, *wbwb*, in south Wales and seems to be evidenced in the form *Treuwbwb* in the diary of William Thomas of Michaelston super Ely in 1766. It may well have been the basis of the error perpetuated by late nineteenth-century antiquaries who linked it with the doubtful *Iupania* (probably, it is thought, for an unidentified *Lupania*) which occurs in the British section of a Latin list of place-names compiled by an anonymous monk of Ravenna at the beginning of the eighth century. It was wrongly read as *Iupu-*

pania, the first two syllables bearing an obvious superficial resemblance to *iwbwb* in the form *Treiwbwb* (Trehwbwb).

It is not certain who was responsible for this correlation. All that will be said here is that the present writer first saw the form *Caerau Treiwbwb* (the plural form of W *caer* having been added, presumably to suggest an early defensive location) in one of Iolo Morganwg's manuscripts in the National Library of Wales. This was certainly seized upon by the estimable editor of the well-known volumes of *Cardiff Records* (1898–1911) who, unfortunately, wrongly identified Iolo's *Caerau Treiwbwb* as the 'ancient encampment' of Caerau on the present western outskirts of Cardiff some three miles distant from Trehwbwb but near enough, it would seem, to assert that the Cardiff Caerau was 'the *Jupupania* of Ptolemy (sic), in Welsh *Tref-iwbwb* "the town of wailing"'.

Tre-os

Travellers in the vicinity of Llan-gan and Llangrallo (Coychurch) near Bridgend continue to be directed on signposts to the hamlet of Tre-oes and not to Tre-os. The latter is the correct form of the name and its etymology is unusual.

The collected forms show that here we have an example of the substitution of W *tre(f)* for the affixed and broadly synonymous E *–ton*, OE *tūn* 'farm, estate' in an original post-Norman name such as those which occur in the Englishries of the Vale of Glamorgan and Gower, like *Trefflemin* for Flemingston, *Trelales* for Laleston etc. The evidence shows that the original name was *Goston* 1525, 1536, 1541, 1559–1724, *Gostonn* 1597, *Coston* 1599, 1531, *Gostson* 1696 etc., and was derived from a ME form of OE *gōs* 'goose' + *ton*. Whether it is to be interpreted as 'a farm where geese were kept' as is probable in Goswick, Northumberland (with OE *wīc* 'dwelling, farm') is not known with certainty. The more likely alternative, since locations in marshy or watery surroundings in England having names containing OE *gōs* are well evidenced, like Gosbeck, Gosbrook, Gosford, Gosforth, Gostrode (with OE *strōd* 'marshy land', as in Stroud), is that Goston was similarly located in the wide and marshy Ewenni valley. Samuel Lewis, in his *Topographical Dictionary* (1833), observed that the soil in the parish of Llan-gan was 'subject to inundation'. Such an area was obviously frequented by wild geese, a feature which was sufficiently conspicuous to have given the holding its name.

The first evidence for the substitution of W *tre(f)* for *-ton* in this case is documented in 1596–1600 in the form *Tress* which can reasonably be assumed to be a scribal error for *Treos* in view of the preponderance of that form thereafter: *Treos (Moor)* 1631, *Velin Treos* 1681–3, *Treos* 1747–8, *Treos* or *Goston* 1833 etc. What makes this form noteworthy, however, is not the substitution of one element for another in a different language but the adoption of the English *gōs*, with no attempt at translation, and its subjection to the normal Welsh process of lenition of the initial consonant in a genitival relationship after *tre(f)*. When the substitution of *tref* for *–ton* was made it was hardly likely that the perpetrators had any inkling of the meaning of *gōs* nor of its linguistic affiliation. It is also clear that the pronunciation of Goston preserved the long value of the vowel in the first syllable, i.e. Gōston, so that the Welsh form is to be pronounced *Tre-ōs* as is confirmed by the testimony of further forms of the name from non-Welsh sources, namely *Treose* in 1799 and 1825 (to rhyme with *rose, close*).

Furthermore, in the eighteenth century the name was subjected to a more deliberate phase of cymricisation, again, it would seem, in ignorance of its true meaning. It also appears to be significant that the first evidence of this change is to be seen in the writings of leading Welsh Methodist personalities of the time. Griffith Jones of Llanddowror in his *Welch Piety* has the form *Tre-Oes* in 1748–9 and Howell Harris apparently wrote *Treves* in his diary in 1750 (a badly formed 'open' *-o-* of Treoes in the manuscript possibly having been read and printed as *-v-*). This accords with the prevailing view that a revival in the fortunes of the Welsh language in the Vale of Glamorgan occurred in the eighteenth century, this being particularly reinforced by the influence and spread of Welsh Methodism, its society groups and associated circulating schools, an influence perpetuated, no doubt, by the activities of the celebrated David Jones in the Llan-gan and Pen-coed area from the late 1760s onwards to the end of the century.

In addition, the re-formation of *Tre-os* as *Tre-oes* may well have been facilitated by the phonological process of reducing an accented diphthong to the value of its dominant vowel component which is a characteristic of spoken Welsh in the south; *-ae-* > *-ā-* as in *maes/mās, cae/cā* etc. Similarly *-oe-* > *-ō-*, as in *troed/trōd, coed/cōd*, and on this pattern the assumption was made that *-os* in *Tre-os* was the reduced form of a basic *oes* and was mistakenly restored as *Tre-oes*. The fact that there is a Welsh common noun *oes* 'lifetime, age, generation, era' may also have been regarded as sufficient justification in an age of religious revivalism when a preoccupation with the spiritual condition of the human lifespan prevailed, notwithstanding its apparent incompatibility as a second element compounded with *tref* in a place-name.

Trerannell : Angelton

A lease of corn tithes granted by the abbot of Margam in 1518 concerns a location *apud penvey et treranell*. The former is Pen-y-fai in the parish of Newcastle, Bridgend, and *treranell* can be confidently identified as the land where the hamlet of Angelton stood at approximately the National Grid reference SS 899816 off the present A4063 road from Bridgend to Maesteg. Nearby is the Glanrhyd Hospital, formerly the Glamorgan County Lunatic Asylum opened for patients in 1864 and built on the land of Angelton Farm which was purchased for the purpose in 1857. Angelton Road and Angelton Green now preserve the name of the vanished hamlet where there was a Great House and at least three other houses, the property of the estate of the Talbots of Hensol in 1636–8, and in 1753–4 one of Griffith Jones's Circulating Schools was being held there.

Early Ordnance Survey maps show a greater number of houses and recent investigations point to the site of a deserted medieval hamlet with house platforms and other remains a little to the west of the marked location of Angelton which may have some significance in this context. The settlement clearly took its name from an original homestead or *tref* which may have been the forerunner of Angelton Farm 1846 but it is not possible at present to ascertain its period of origin.

The original Welsh form of the name is recorded as *Treranell* 1518, 1604–5, *Treranol* 1584–5, *Trerannell*, *Trerannell's Greene* 1623, *Tre Annel* 1636, *Tre yr Annel* 1638, and in 1618 there occurs the crucial form *Tre Angel*. By the time of Griffith Jones's *Welch Piety* we have *Angel Town* 1753–4, and thereafter *Angelstown* 1758, *Angel Town* 1779, 1824, *Angelston* 1785, *Angelton* 1814, 1833 etc. with the E *-ton* being substituted for W *tref* as in so many other instances.

The key form which identifies the original *Trerannell* as Angelton is the 1618 *Tre Angel*, and it should be appreciated that the second element in this form would have been pronounced in the Welsh, and not the English manner. It has nothing to do with what the OED defines as 'an order of spiritual beings superior to man in power and intelligence', although popular etymology may well have made that connection once the anglicised form in *angel-* had become current. Two stages of development can be detected in the forms listed above. Those of 1584–5, 1636 and 1638 show the anglicising tendency to pronounce the Welsh fricative *-ll* as *-l*, *-annell* > *-annel*, to which can be added the nasalisation of *-nn-* to *-ng-* > *angel*, which can be compared with a similar process in *Yrannell* > *Yrangell*, a lost stream name now preserved in Blaenyrangell, Pembs.

The full form of the original second element *-annell*, or more correctly *-rannell*, was *ariannell*, a W diminutive form in *-ell* which is related to W *arian* 'silver' and is also to be seen in the form *arannell* by loss of the consonantal *-i-* common in south Wales spoken forms. This is noted by R.J. Thomas as the name of at least eleven streams and rivers in Wales and was probably descriptive of the 'silvery, bright, foamy' flow of their waters, but the form *Ar(i)annell* is also attested as a Welsh personal name, this being the more likely interpretation of the element compounded with *tref* in *Tref Arannell* (to give it its likely original form) to indicate possession. For instance, a variant form of the personal name (by ultimate *-i-* affection) is *Eiriannell* which occurs in Pentre Eiriannell in Anglesey. In the pers.n., the suffix *-ell* still has a diminutive function to convey a hypocoristic or 'pet' sense signifying 'the small bright, silvery or fair-headed one'.

Finally, a brief comment on the form *Trerannell*. Having confidently postulated *Tref Arannell* as the original form, we can be clear that *Trerannell* could have evolved in spoken forms by the common reduction of *tref* to *tre* and the loss of the unaccented leading vowel of the second element *arannell*: *Tref Arannell* > *Tre-arannell* > *Tre(a)rannell* > *Trerannell*. However, whether this is in fact what occurred is open to question and cannot be proved without the further evidence of forms earlier than 1518. Another possibility is that the unaccented *a-* of the second element could first have become less distinct in the vernacular until it resembled the 'dark' sound of W y, so that *aránnell* > **yrannell*, and by confusion with the full form of the W def.art. *yr* an erroneous division into two elements *yr annell* could have occurred. This is confirmed by the fact that the name does occur simply as *Annell* in other places in Wales associated with streams which have the same name, such as Aber-annell, Blaen-annell and Esgair Annell with which the forms *Tre yr Annel(l)* 1638 and *Tre Annel(l)* 1636 of the name under discussion can be compared.

Further, a spoken form like **yrannell* could be erroneously divided into the vocalic form of the W def.art. *y* + what would then appear to be the element *-rannell*, i.e. *y rannell*. This may well have happened in an example of Arannell which occurs specifically as a stream name in Glamorgan. This is recorded as *Ranelh, Ranel* 12c, *Raneth* (leg. *Ranelh*) 1517, and is the brook whose name is well documented as the eastern limit of a tract of land having its western limit the stream Ffrwdwyllt 'the turbulent stream' (*terram qui jacet inter Ranel et Frudel* 1149–83) which flows into the Afan estuary near Tai-bach. This Arannell rises on the slopes of Mynydd Margam and flows down Cwm y Brombil (p. 23). It bears the name *Arnallt Brook* on early Ordnance Survey maps and gave the name *Arnallt* to a property (a 'freehold house' in 1846) which stood on its bank at approximately SS 783875 in an area now built up. Another form of the name also appears in the modern Rhanallt Street at SS 779881.

One further development which did not survive is seen in the form *Danielston or Trerannel* 1703, *Tre Daniel* 1716, which is difficult to account for other than by consignment to the realm of fanciful reconstruction. The isolated nature of these eighteenth-century forms can hardly sustain an argument, otherwise uncorroborated, for the connection of a person named Daniel with the area.

Tumbledown

The A48 road climbs gradually westwards from Culverhouse Cross on the western outskirts of Cardiff to reach the higher expanse of St Lythan's Down on which the BBC television mast stands near the hamlet of The Downs. This well-known hill is known locally as the Tumble Hill, a name which has been the subject of much speculation and some onomastic invention, for the fact is that there is little surviving evidence in documentation available to aid confident interpretation.

A charter of 1186–91 granting twelve acres of land to Margam Abbey states that it was located on *Turbernesdune* and it has been accepted that this refers to the general area of St Lythan's Down, the second element of the name being OE *dūn*, possibly in its original sense of 'low hill, upland expanse', the modern reflex being *down* which is used in field and minor names for landscape features. A reasonable case has been made for a personal name, *Þorbjorn*, ultimately (if not directly) of Scandinavian origin, as the first element to indicate possession or other association, 'Thorbjorn's down'. It has also been claimed that the modern Tumble is a shortened version of an original Tumbledown and that the twelfth-century *Turbernesdune* was its precursor, presumably on the basis of a superficial similarity of form for there are no intervening dated forms available to corroborate this as far as is known. It is a suggestion which is by no means implausible, although difficult to prove.

It may be relevant to observe that the form Tumbledown occurs in England in the combination Tumbledown Dick as an inn sign which depicts a jovial red-coated old farmer figure with a bottle or pot in his hand falling out of his chair in an inebriated condition. No doubt by the present time the personal name Dick can stand in this context for any person, as in the phrase 'Tom, Dick and Harry', but in the years closely following 1659 such a designation had a more specific significance, particularly in areas which were royalist in sympathy, as a disparaging reference to Richard Cromwell, the son of the redoubtable Oliver and his successor as Lord Protector but only for eight months before his downfall on the restoration of Charles II. The allusion is common in satirical writings after 1660.

Was there an inn in the vicinity of the Tumble Hill whose name was Tumbledown Dick and given in this manner? The remarkable coincidence of similarity of form with *Turbernesdune* has led to a suggestion that Tumbledown was a so-called 'corruption' of the earlier name, but this is unlikely if, as the lack of dated evidence suggests, *Turbernesdune* was a name long forgotten in the locality. The same consideration makes the idea that a measure of popular etymology may have been responsible for the emergence of Tumbledown from the earlier form equally unlikely.

On the Tumble, about half way up the slope and slightly off the present A48 stands the Trehern Arms so named after the Traherne family of Coedriglan nearby (p. 43). Before the present wide road was developed an earlier and much steeper road or track ran almost parallel with it and behind the hostelry which a nineteenth-century local historian called 'an Inn called Tumble Down Dick', echoed by another in the 1930s as 'the Tumble Inn'. It is not known on what authority these statements were made but it would seem that the area of the Downs was not immune to military movement during the Civil War prior to the battle of St Fagans in 1648, one letter from an active group of royalist gentry to Major-General Laugharne being dated from St Lythan's Down in 1647. Also, it is reasonably clear that the inn would not have acquired its present name until the latter half of the eighteenth century at the earliest when Llewelyn Traherne, formerly of Castellau, Llantrisant, inherited Coedriglan.

Of course, a location on a steep hill was not necessarily a prerequisite for naming an inn the Tumbledown Dick, but possibly a popular notion that a steep hill was conducive to a fall or tumble may have facilitated such a designation in the naming process. The name can be compared with that of the village of Tumble (Y Tymbl) which overlooks the upper reaches of Cwm Gwendraeth on a slope near Crosshands, Carms. There is some record of a public house called the Tumble Inn in the vicinity. Could this have been originally Tumbledown Dick and at what date? It would appear that such a name can be of comparatively recent provenance, as was the case with another Tumbledown Dick which no longer stands between Aberafan and Cwmafan, near Port Talbot. This was at one time a farmhouse bearing the name of *Tyle Cnwc* (W *tyle* 'slope, ascent, hill' + *cnwc* 'hillock, knoll') converted into a public house and named Tumble Down Dick as late as *c.* 1824.

Twynbwmbegan

On the left of the road which leads from Culverhouse Cross to Wenvoe (Gwenfô) and near the Vale of Glamorgan Council's storehouse the land rises to form a substantial hillock. Under its flank David Davies of Llandinam's railway tunnelled its way to Barry and its summit commands a view of a wide sweep of the Vale from the Severn to Craig Llanishen to the north-east. This is Twynbwmbegan, with what seems to be an unusual second element after W *twyn* 'hill, tump, knoll, rising ground' etc.

Although recorded forms of the name are not plentiful there seems to be no reason to suspect the validity of their mainly eighteenth-century sources: *Broombeacon* 1784–5, *Broombegan* 1787 and *Bwnbegan* 1798. These indicate that *twyn* must have been a relatively late and tautologous addition to what was originally an English name of two elements, namely *broom*, OE *brōm*, the plant name, and in that sense here to signify the predominant feature of the vegetation at one time (although the word seems originally to have denoted a thorny bush or scrub) cf. Brombil (p. 23) + *beacon*, OE *bēcun* 'sign, signal' used for a place where a beacon-fire could be made for signalling, usually an elevated site, a name found commonly in England in the combination Beacon Hill. Although *beacon* can also be found combined with other elements and having a range of meanings it is apparent that it came to be used on its own for a hill or height as in names like the Brecon Beacons (the Welsh name being Bannau Brycheiniog where *bannau* is the plural form of W *ban* 'hill, height, crest' etc.) or Dunkery Beacon on Exmoor in Somerset.

An unusual loss of –*r*- in the first element *broom-* > *boom-* has occurred, represented by the cymricised *bwm-* > *bwn-* in common speech by the end of the nineteenth century, a change similar to the –*m* > -*n* in the Welsh name of the bittern, W *bwm(p)* (imitative of the bird's 'boom') > *bwn*, otherwise *aderyn y bwn*. The second element *beacon* has been borrowed into Welsh as *bigwn, begwn*. In reply to his Parochial Queries at the beginning of the eighteenth century the antiquary and philologist Edward Lhuyd was told that there was a *bigwn* in the parish of Bugeildy, Radnor, which is the present Beacon Hill. GPC notes that the form *y Begwns* has been recorded as a name for the Brecon Beacons, also that *pegwn* occurs as a variant form in the place-names Pegwn Bach and Pegwn Mawr in the parish of Llandinam, Mont., with *pecwn* the colloquial form used in west Glamorgan for the summit of a hill.

Twyncyn

This name, which is a compound of W *twyn* 'knoll, hill' + the W diminutive suffix -*cyn*, borrowed from E -*kin*, as in a pair like *bryn, bryncyn*, is to be found quite commonly throughout Wales, but the Twyncyn noted here, that which occurs in Dinas Powys, is not directly so named, although it could be accepted as a suitable description the land which rises gently to the north-west of the village.

It cannot be doubted that it is to this land that an early sixteenth-century documented reference applies in the form *Tomkin Smith's*

Lande, and thereafter *The Tompkin* 1765, *The Tomkin* 1784 and *Tomkin Land* 1785. In the vicinity today stands Coed Twyncyn 'Twyncyn Wood' which was *Tomkins Wood* 1685, 1693, *Tomkin Wood* 1771, 1796, *Great Tomkin's Wood* 1783 etc.

It would appear, therefore, that Twyncyn in this case is a late form adopted in order to give meaning to a local spoken form of *Tom(p)kin* (for which an oral form broadly similar to *Tonkin*, *Twnkin*, was obtained) which was otherwise devoid of meaning to a later generation, possibly towards the end of the nineteenth century and, perhaps, aided by the fact that it referred to rising ground. As far as is known, this is the only example of its kind.

What is significant, of course, is that this is a Welsh re-formation of an original English name and that it occurs in an area in which the English language had long predominated. It could be regarded as indicative of the known revival in the fortunes of the Welsh language in the Vale of Glamorgan and traceable to the eighteenth and early nineteenth centuries.

The original name perpetuates the name or surname of a person who held the land at one time, of whom we have no further record, Tom(p)kin being a diminutive form of Tom (Thomas), cf. Wilkin, Watkin etc., and the surviving evidence of related name-forms shows that a variant form of the personal name was in existence. In 1784 a *Tomkin Lane* is recorded. Earlier, in 1773, it was *Tinkin Lane*. It may well be possible, therefore, that this is the personal name which is an element in the name of the well-known Stone Age burial site in the parish of St Nicholas, *Tinkinswood*.

Uchelolau : Highlight

At present, Highlight appears on the map as the name of a farmhouse on the north-facing slopes of a low ridge of land looking down on the valley of the Waycock stream to the north of Barry. The surrounding area is now recognised as the site of a deserted medieval village which was depopulated by the beginning of the sixteenth century. Originating as the name of a sub-manor of Dinas Powys in the twelfth century, it was also a parish until 1898 when it became an extra-parochial district of the neighbouring parish of Wenvoe, its church having been abandoned *c.* 1570 but commemorated in the name of the nearby Coed y Capel.

The first date of the appearance of the English form of the name seen hitherto is *Highlight c.* 1558–1603. Earlier, the original name had the Welsh form Uchelolau: *Hukheloleu* 1254, *Yughawley* 1538–44, *Uchelole* 1550–1600, *Yeholey* 1566, *Ychol(l)ey* 1566, *Ychelloley* 1567 etc., with very corrupt documented forms *c.* 1291, *c.* 1348, 1373–4, 1423–4, the name having presented considerable difficulties for non-Welsh scribes. The English form Highlight is clearly a literal translation of the Welsh form, based on the assumption that the the two elements in the name are W *uchel* 'high, elevated' and the initially lenited form, *olau*, of W *golau* 'light'. This has prompted references either to the lights used in medieval churches for different purposes, before altars, images etc., or to lights shining outwards and visible in relation to the surrounding countryside in attempts to explain the meaning of the name.

A more convincing interpretation of the second element *olau* (accepting W *uchel* 'high, elevated' as the first) is that it is a W plural form in *-au* of the n.m. *ôl* 'track, course, path' as in the common phrases *ar fy ôl, ar ei ôl* 'after me, after him', i.e. 'in my track, in his track'. The pl. form *olau*, MW *oleu*, is well-attested in a collective singular sense in medieval Welsh texts between the thirteenth and sixteenth centuries, e.g. *a gwelet oleu y meirch awnaethant* (13c) 'and they observed the track(s) of the horses', such a track or trace ultimately constituting a 'path, way' which, in conjunction with W *uchel* used in the elliptical sense of 'high ground', suggests an elevated path, track or way, perhaps a 'ridgeway'. As noted above, the site of the village is on a low ridge of land, and the evidence adduced in the Royal Commission on Ancient and Historical Monuments' Inventory of Non-Defensive Medieval Secular Monuments in Glamorgan is that 'the ridge is traversed by an ancient and deeply worn trackway' which runs down past the site of the church.

This interpretation is supported by the existence of other examples of the name which occur in Glamorgan and their locations. A farmstead near Ewenni was *Eghelloley* 1611, *Ychylola* 1799, *Uchel-Oleu* 1886, but *Highlight* 1846 and may be related to an early road or track in that vicinity, whilst on each side of a narrow way which runs northwards from Pen-coed, near Bridgend, through Rhiw'rceiliog and up the southern slope of Mynydd-y-gaer over the eastern extension of which it now runs unfenced to drop down into Cwm Ogwr Fach stood two farmsteads, no longer in existence though their sites are visible, which bore the name. On an elevated position on the sharp southern ascent stood *Uchel Olaf Uchaf* (sic) on the first edition of the Ordnance Survey one-inch map of 1833, and *Uchel Olaf Isaf*. Before the sub-division of the property into two holdings (*isaf* 'lower', *uchaf* 'higher') the single holding was *Ychylola* 1799, *Lecholola* 1801, *Uwchalola* 1846. The road appears to have been an important line of communication, probably the only way across the considerable extent of high ground of which the south-eastern spur of Mynydd-y-gaer is the most prominent feature and thought on good authority to have been in use in the early Christian period.

Further, as noted in discussing Heol-y-march (p. 98) the W element *olau* also forms the second element of the name of the mansion of Rheola on the western side of the Vale of Neath. Its original form was *Hir-olau*, with W *hir* 'long' as first element + *olau* > *ole* > *ola* in common useage, 'long track, way': *Hirrole* 1295, *Hyrolle* 1295-6, *Hirolle* 1376, *(Mellin) Hire Oley* 1617, *Hireoley* 1657 etc., *Rheola* 1763, *Rhyola Farm* 1812 etc. The location is below the ridge of Hirfynydd (*hir* + W *mynydd* 'mountain') along which runs the known straight two-mile extent of the so-called Sarn Helen, the Roman road running north-east to Coelbren, and from where it enters the modern Rheola Forest. Whether this road is, in fact, the *hir-olau* of the place-name cannot be proved for lack of earlier forms but the coincidence of position in relation to the road is noteworthy, and the fact that the name is evidenced in the thirteenth century proves that the reference is not to some modern construction.

Verville

A low ridge of land projects into the triangle formed by the junction of the rivers Ogwr (Ogmore) and Ewenni near Ogmore Castle and not far from their estuary at Aberogwr (Ogmore by Sea). Standing prominently upon it is the farm which bears this name.

It does not have the appearance of a Welsh name, indeed, it is not Welsh, but the temptation to see French influence in the final element *-ville* should be resisted, as this is merely an attempt to represent a form of pronunciation which, it may be granted, is not far off the mark. For it is an English name, the basic form of which is well attested in English field names (in Wales as well as in England), and is compounded of the elements *fore* 'in front of, before', sometimes implying a feature which is prominent or juts forward + *field*. The location in the present instance on the higher ground in wet and marshy surroundings is such that can be aptly described in such terms, but certainly not in the sense suggested in two recorded instances, namely *Varfrell alias Fayre ville* 1627–8 and *Fairfield alias Vervil* 1807. It is otherwise recorded as *Vervill* 1670, *Vervil fach* 1754, *Little Vervil farm* 1774, 1787.

Concerning the use of the element fore in place-names in Glamorgan, Foreland (*Forelands Close* 15-16c) is recorded in the parish of Lecwydd (Leckwith), Forty (with OE *forð* of similar meaning to *fore*) in St Brides super Ely (p. 75), and a name corresponding to Verville in the form *The Vorfill* 1716, *The Vorvill* 1774, in St Andrews parish. These forms, as has been noted previously, exhibit dialectal features which are those of the south and south-western counties of England brought to the Vale of Glamorgan and Gower in particular by immigrants from those areas over the years. One distinct feature is the voicing of initial *f-* to *v-*, so that field becomes *vield* and *viel(d)* with loss of the final consonant which is represented by scribes in documentary forms as *ville*, cf. Cwrt-y-fil (pp. 59–60).

It is this influence which makes clear the interpretation of names in the Vale and Gower such as Vurlong, Tŷ Verlon (*furlong*), Vishwell, Welvord, Vershill (*furze*), Vernel (*Vernell* 1543–4, 1739, *Vernhill* 1650, *Vernel* 1742, 1764 etc., *fern* + *hill*) near Casllwchwr (Loughor), and the recording of Fernhill Farm in Rhosili as *Vernals* in 1820.

Watford

This was originally the name of a holding of land on the northern slopes of Caerffili Mountain or Common where it merges into Twyn Garwa and close to a stream, the Nant Ddu, later subdivided into two properties known as Watford Fawr (later Plas Watford) and Watford Fach. The former was the home of Thomas Price, one of the co-partners who took over a lease of mineral rights on the bank of the Dowlais brook in Merthyr Tudful in 1759 to found the great ironworks in that location. Perhaps he is better known for having granted leases to the Independents of the Caerffili area to build a meeting-house, the well-known Watford Chapel, on a portion of his land in 1739 and to have been host to the early Welsh and English Methodist leaders whose first joint association was held at Watford in 1743.

Watford has the appearance of an English name in common with the better known Watford in Hertfordshire and Northamptonshire. In both cases, these have OE *wāth* 'hunting, chase' as their first element + the common *ford* to give the sense 'a ford used when hunting'. Perhaps Howell Harris, the Welsh Methodist leader, has himself contributed to the impression that the name was of English origin although his implied derivation was different when he wrote *Waterford* in his correspondence 1739–44, possibly under the impression that it signified a ford on the Nant Ddu. However, Harris is not consistent, for he uses the form *Bodford* 1739 as well as *Vodford* (= *Fodford*), an initially lenited form after a lost W def.art., but this form seems to be nearer the mark as it is attested in other recorded forms. This may also be the form that George Yates was trying to convey in his 1799 map of Glamorgan with *Uoltvor*, the medial -*d*- being unvoiced before the voiceless consonant -*ff*- > -*tff*-. The place-name Bodffordd, Anglesey, is always pronounced *Botffordd*.

Boddfordd occurs in 1738–9, probably written by a non-Welsh scribe, where the medial *-dd-* is probably to be pronounced like an English *-dd-* in *middle, saddle* etc., and with the English *-f-* for the digraph Welsh *-ff-*. In the estate papers of the family of Turberville of Ewenni the form *Tir y Bedfordd* 1717 is recorded, where *bed-* appears as a variant form of *bod-*, as in names like Bedwenarth and Bedwellte, confirmed by *Tyr y Botffordd otherwise Watford* 1835. Again it is *Tyr y boteford* in the earl of Pembroke's estate survey of 1570 and two earlier forms are combined with E *way*, OE *weg* 'track, way', *Wotfordesweye* 1313 and *Botfordwey* 1314–15. It has also been maintained that locally Plas Watford was usually referred to as *Y Fotffordd Fawr*. In view of this evidence it may be worth considering whether the name has a similar origin to Bodffordd, Anglesey, noted above which my colleague Tomos Roberts derives from W *bod* 'dwelling, abode' + E *ford*.

W *bod* in this sense is a common element in place-names in north Wales, over a hundred examples having been recorded in Anglesey alone, but it was not used frequently in the south, Bedwenarth and Bedwellte being two examples noted above, a lost *Bodesgob* on Cefnsaeson near Neath and an uncertain *bod yssell* 1561–2, *bod isell* 1630 in the parish of Eglwysilan being also among them, whatever remains to be said for the still uncertain Bodringallt and Bedlinog.

The second element which appears as W *ffordd* would mean 'road' in its present sense if it was of comparatively recent provenance but it is originally a loanword from OE *ford* in its early sense of 'way, passage, route' and has been in the language since at least the twelfth century, if not earlier. Some place-names in England which have *–ford* as the second element, as Margaret Gelling notes, are either on long-distance medieval routes or routes along which neighbouring villages communicated with one another rather than containing specific references to river-crossings as they would in more recent times. The reference could also be to causeways or a passage through wet ground in other English names. Conversely, it is conceivable that examples exist of Welsh place-names containing *ffordd* in the sense of 'ford'.

In the *Liber Landavensis* the starting point of a recital of locations on the boundary of the parish of *Llanwynwarwy* (colloquially *Llanwarw*, now Wonastow, Mon.) is named *ir ford artrodi* (= *y ffordd ar Droddi* 'the ford on the Troddi', Troddi being the name of the stream now often rendered Trothy). This becomes *ir rit artrodi* (= *y rhyd ar Droddi*, using the alternative W *rhyd* 'ford') at the end of the list where the name of the starting point is repeated. It is also interesting to note that the Welsh forms of some English names in Wales and the borders have *-ffordd* for E *-ford*: Hereford, W *Henffordd*, Haverford(west), W *Hwlffordd*, Pulford, W *Pwlffordd*, Cheshire, and Whitford, W *Chwitffordd*, Flintshire.

It is possible, therefore, that Bodffordd, Anglesey, means 'the dwelling by the ford' as there were three fords in the vicinity, one on the outskirts of the village, rather than 'the dwelling by the road'. If this is the original form of Watford the same alternative senses could be applied. The uncertainty remains, however, the main consideration being that the earliest recorded form of the name evidenced to date is not helpful. It is *Watford* 1307.

Y Werfa

The original farmstead of this name on the western slopes of Mynydd Aberdâr (Aberdare Mountain) below the ridge of Twynywerfa (which preserves the name) and overlooking the site of the old Abernant ironworks no longer stands. Its land, specified as *Tyr y Werva*, was leased for industrial development in 1801, with permission to erect a blast furnace on the land of the neighbouring Abernant y Wenallt farm. I am told by my correspondent, Edward G. Williams, whose family occupied the old farm of Y Werfa from 1866 to 1968 and who was a descendant of previous owners in the eighteenth century, that the present Werfa House, which stands lower down the hillside, was built *c*. 1886, some of the stonework having come from the old farmhouse of which barely a trace is now visible, the site being closely overgrown with trees.

The earliest recorded form of the name seen hitherto is *Tyr y Werva* 1602 and the form does not vary to any great extent in documentation thereafter except for the obvious v/f variation, *werfa/werva*. It is clearly a name of two elements, the second being the common Welsh ending *-fa*, the lenited form of *-ma*, originally 'a plain, a field' as the first element in names like Machynlleth, Mathafarn, Machen etc., later 'place, spot' as a suffix in words like *porfa* 'pasture', *rhodfa* 'walk, avenue' etc. It occurs in the latter capacity here with what appears to be by the present time an unusual first element although it occurs in other names in Glamorgan: Y Werfa Ddu, Llantrisant, Maesywerfa near Bryncethin, Rhiw y werfa

(1790), Hirwaun, and also in Mon. and Cards.

In essence, this is a combination of the Welsh prefix *go-*, which moderates the sense of the word that follows in combination (as in the well-known Welsh phrase *go lew* 'fair, fairly good' < *go* + *glew* 'valiant, steadfast, splendid') but here with the W adj. *oer* 'cold' > *go-oer* 'cool, shady', probably with reference to pasture for cattle and sheep shaded from the heat of the sun or in the shadow of a grove or plantation. In the form *göer* the vowel combination *-o-o-* after the initial *g-* resulted in a coalescence which produced a long *–ŵ-* to give the variant colloquial form *gŵer* which, as is evidenced in GPC, is heard in common speech in Brecs. and Carms. as well as areas of Glamorgan, sometimes in the form *gywer* (the name of a piece of ground near the parish church of Leckwith must have been pronounced in that way for it to be documented as *The Gower* in 1832).

Further liberties were taken with the quality of the vowel in the vernacular in some regions, as in the Aman valley where the *-e-* becomes accented and long to give *gwêr* which could be the basis of *gwerfa* (< *göer-fa*) this, in turn, having a variant form *gwyrfa* which can be heard in other areas of Glamorgan and in Mon. The name discussed here appears as *Wyrfa*, *Wyrva* in nineteenth-century documentation.

Wharton Street : Heol-y-cawl

Wharton Street is the name by which a street which runs into St Mary Street from The Hayes past Howell's store in Cardiff is usually called nowadays and may well be the *Wottonstrete* recorded in 1492. If this is so, that is the earliest form of the name evidenced hitherto. Far more certain as early forms, however, are *Worton strete* 1549, 1551, 1560, *Werton Strete* 1549, *Wrotton-streete* 1557, 1563, *Wrotton Lane* 1557, *Woorton strete* 1560, *Wortton Strete* 1563, *Worten Streete* 1616 etc., the main element being the OE compound *wyrt-tūn*, which has OE *wyrt* 'a plant, a vegetable', now *-wort* which occurs in plant names like *colewort*, *liverwort*, but once used for any plant of the brassica or cabbage kind. Thus *wyrt-tūn* had the meaning 'vegetable or kitchen garden' (with OE *tūn* here in the sense of 'enclosure'). In this case, it is relevant to consider further the word *colewort*.

E *colewort* is a compound of two broadly synonymous elements, the first being OE *cawel*, *cāl*, from Lat *caulis* used for various species of brassica and is, of course seen in the form *kale* and the first element of *cauliflower*. It is from the Lat form that W *cawl* also comes, so that it is not entirely surprising to find a Welsh equivalent to Wharton Street, namely Heol-y-cawl: *Hewle y Cawle* 1682, *Hewl Cawl* 1737, confirmed as *Worton Street commonly called Houle Cawle* 1768. The W form is *heol* 'street, road, way' + the def.art. *y* + *cawl*. The original reference here may have been to burgess garden enclosures or to a wet and muddy lane where wild cabbage, perhaps smallage, grew, as in several instances of Heol-y-cawl elsewhere, in the parish of Llanilltud Faerdre and St Andrews etc. This is also the *cawl* which appears in the name *Porth-y-cawl*, now Porthcawl (having an alternative *Pwll Cawl* in 1825) with W *porth* 'harbour, haven' + *cawl*, a reference to the abundance of sea-kale or wild cabbage which was to be found on pebble ridges and amongst the coarse grasses of the Glamorgan coast.

However, that is not the end of the story. On John Speed's map of Cardiff in 1610 the name of the street appears as *Porag Stret*. By 1755 it has become *Pottage or Porridge lane*, and *a lane called Porridge Lane* 1815. The reason for this is that W *cawl* had developed an extended meaning in the course of time from its original designation of a particular plant to that of a dish of vegetables, alone or with meat, boiled to produce a broth, soup or pottage, which it denotes at present. Such is the explanation for Heol-y-cawl being literally translated as Pottage or Porridge lane (*porridge* being a variant form of *pottage*).

Further, in a guidebook of 1829 there appears another form of the name, *Broth Lane*, defined as 'an old name for Wharton Street, Porridge Lane or Hewl y Cawl', the wheel having turned full circle.

Whitchurch : Yr Eglwys Newydd

The question most often asked about this name is why the Welsh form, *eglwys* 'church' and *newydd* 'new', is not a literal translation of the English form, which is recorded as *Witechurche* 13c., *Whytechurch* 1376, *Whitchurche* 1385, 1440, *Whittechurche* 1443, *Whytchurche* 1492, and so on until the present day? The short answer is that it is not an incorrect translation of the English form but a separate and partly unrelated Welsh form.

In the early Middle Ages a chapelry of Llandaf was founded on the banks of a pronounced curve of the river Taf, its location being called *Ystum Taf* (W *ystum* 'bend, curve') which appears in a twelfth-century reference to the chapel as *Capella de Stuntaf*, 1126. This is no longer in existence but it may well have been situated in the vicinity of the old Melingriffith tinplate works, the name Ystum Taf being now resurrected by the Welsh-speaking community as that of the surrounding urban growth, officially known as Llandaff North.

In thirteenth- and fourteenth-century Latin documentation the chapel was referred to as *Album Monasterium*. The synonymous English *Whitminster* and the Norman-French *Blaunk Moustier* occur in 1315, *Blankmoster* 1322 (*minster* and *mou(s)tier*, like W *mystwyr*, being derived forms of Latin *monasterium*, the interpretation of which must be influenced not by thoughts of an ornate stone-built late medieval edifice, but rather of an early religious 'cell'). This is the basis of Whitchurch.

By the middle of the thirteenth century there had come into existence the demesne manor of Whitchurch which was probably more or less territorially co-extensive with the modern parish. Its administrative centre was the small castle of Treoda on its motte near which stood both the later substantial residence of that name and another chapel or church which continued in the similar status of a chapelry of the cathedral church of Llandaf as the earlier *Album Monasterium*.

It is to this church that John Leland refers in 1536–9 as *Egluis Newith*, and in a well-known list of Welsh parishes *c*. 1566 *yr eglwys newydd* confirms the authenticity of Leland's form. Indeed, *Newchurch* 1472, *Newchurche* 1600–7 is attested as a form which obviously did not survive or supplant the earlier Whitchurch. But neither is there recorded a Welsh equivalent of Whitchurch, which would have been *(Yr) Eglwys Wen*. By 1845 the modern parish of Whitchurch had been created and a new parish church, that of St Mary's, opened in 1885. In the meantime, the fabric of the older church (Leland's 'new' church, *Egluis Newith*) had deteriorated to such an extent that it was demolished in 1904 and its foundations consolidated as a public amenity in 1973. Its preserved site is located in the present Old Church Road.

It is the name of the early chapel of *Album Monasterium* or *Whitminster* which became Whitchurch by the thirteenth century, on the one hand, and the Welsh name of the later medieval *(Yr) Eglwys Newydd* on the other, which have survived as the names of the modern parish, later (in 1967) to be included within the boundary of the city of Cardiff as one of its populous suburbs.

The significance of the first element *white* in this particular case is open to conjecture as there is no hard evidence to suggest an answer, but it would seem to be a significant fact that the Latin *album monasterium* and Norman-French *blancmoustier* are characteristic and well-attested documentary renderings of this, the commonest *-church* compound in English place-names, according to Dr Margaret Gelling. The two main possibilities are (a) the obvious reference to colour arising from the practice of white-washing church walls, as in the reference in the *Historia* of Gruffudd ap Cynan to the *eglwysseu kalcheit* 'the lime-washed churches' of Gwynedd, or (b) a much favoured interpretation in most examples in England, namely a reference to the lighter colour of a stone-built church or chapel as opposed to one built of wood. Bede's explanation in his *History of the English Church and People* (AD 731), that *Candida Casa* 'white house' in Galloway was so called because the church was built of stone can be recalled, this being echoed in the later Old English translation of the work as *(aet) Hwītan Ærne* (OE *hwītan*, the oblique case of *hwīt* 'white' + OE *ærn* 'house, building').

Wormshead : Ynysweryn

There was once a tendency to regard the name of this well-known headland or rocky promontory which extends into the sea from the southern point of Rhosili Bay in Gower as being of Scandinavian provenance, a comparison being made with the name of the equally well-known Great Orme's Head at Llandudno, Caerns., which is accepted as having ON *ǫrmr* 'snake' as its main element. However, it was effectively argued by Professor Melville Richards that where authentic names of Scandinavian origin in Wales are concerned, particularly those which have Welsh alternatives like Swansea and Abertawe, Fishguard and Abergwaun, and for that matter the Great Orme and its earlier Welsh counterpart, *Cyngreawdr Fynydd* (later *Penygogarth*), those Scandinavian forms have neither a phonological nor a semantic relationship with the Welsh names. This is the most telling consideration against accepting Wormshead as having ON *ǫrmr* as its first element. Neither can the form be evidenced until the fifteenth century, although it may be granted that several names which are likely to have been of Scandinavian provenance are accepted as such not because they have survived in early documentation but on comparative linguistic grounds.

The headland is named, in common with many others, as well as small islands, rocks and reefs, according to the shape of an animal which it resembles in outline when viewed from the sea. The projecting limestone formations at this location seem to have suggested the undulations of the back of a serpent-like creature. The antiquary B.H. Malkin in 1807 affirmed that 'It derives its name from a fancy of the sailors who imagine it to resemble a worm, crawling with its head erect ... '. Further, since the tide at full flow isolates the headland from the mainland it is called *insula Wormyshede* in a fifteenth-century itinerary and by Rice Merrick of Cottrel in 1578, 'Wormshead, a little Iland, a place notorious to seamen'.

This tends to suggest the use of E *worm*, OE *wyrm* + *head* as elements in the name, the earlier meaning of OE *wyrm* being 'a reptile, a snake', for in addition, in a fourteenth-century Latin version of the life of the saint Cenydd, the headland is called by the Welsh name of *Henisweryn*, in modern orthography Ynysweryn. This is W *ynys* 'island' + *gweryn*, a word which has lost currency in ModW but has a meaning similar to that of OE *wyrm*, the forms of the two words being possibly related and can be compared with Lat. *vermis* 'worm', one suggestion being that they derive from a common Celt. *$u̯er-in$-* from a root-form *$u̯er$-* 'to turn, twist, entwine' (GPC). It is not inconceivable that the two forms of the names Ynysweryn and Wormshead are renderings, the one of the other, but the lack of earlier recorded forms makes it difficult to establish which may be the older of the two.

Such are the considerations which would seem to tilt the balance against a Scandinavian provenance, the survival of ON *ǫrmr* in the form which it takes in the name of the Great Orme making that name a more likely candidate for such a derivation. For there are yet more alternate Welsh forms used as names for Wormshead. A modern and recently more frequently used term is Penrhyn Gŵyr 'the Gower Promontory' which can be justified as the name of the most westerly point in Glamorgan, having W *Gŵyr* 'Gower' as the qualifying element. However, the Welsh form which is usually cited as being the formal alternative is Penypyrod, but it has an artificial look about it based as it seems to be on an uncertain element. The W *pen* 'head' is, of course, normal, but *pyrod* is difficult to define if, indeed, it can be accepted. Dr B.G. Charles suggested that it was an affected, perhaps plural, form based on a W *pwr* 'worm', a word which does not even find a place in GPC. Finally, it is Rice Merrick again who provides another possibility. In some notes on the lordship of Gower he uses the form which appears in one transcript as *Pen-y-prys* where the second element seems to be W *prys(g)* 'copse, grove' , an unlikely feature in such an inhospitable environment. Brian James, however, in his authoritative version of Merrick's *Morganiae Archaiographia*, prefers to read *Pen-y-pryf*. This is very probably the correct form, the final element here being W *pryf*, a word which has acquired a number of meanings, the usual modern sense being 'insect, fly', but it is recorded since the thirteenth century in the sense of 'reptile, serpent, snake' thus appearing to be the author's literal and accurate translation of Wormshead.

Yniston

The farm known by this name stands prominently on a bluff overlooking Leckwith Moors which the road from Cardiff to Llandough and Barry ascends after crossing the Ely river near the site of the old Leckwith Bridge. The name has caused some difficulty over its interpretation in the past probably because of the failure to appreciate that it is a name which has changed its form because of a strong Welsh influence in common speech and that in an area not noted for the strength of its Welsh-speaking population. The map form today is *Ynyston Farm*, suggesting that it is a compound of W *ynys* 'island, water-meadow' + W *ton* 'unploughed land, lay-land, grassland' as in Tongwynlais, Tonpentre etc., whereas, in fact, it is neither of these, nor is it *Islwyn*, as has been suggested, with W *is* 'below' and *llwyn* 'grove, copse, bush', nor *Istwyn* (*Is-Twyn* 1833 on the first edition of the Ordnance Survey one-inch map) with W *twyn* 'hillock, knoll', a form possibly influenced by the *Nishtwyn* of a Bute estate map in 1773.

Towards the end of the fifteenth century the form was *Eston*, later *Easton* 1660–80, *Easttown* 1714, which is self-explanatory and refers to the *ton* (OE *tūn*) 'homestead, farmstead' on the eastern rim of the parish (this being certainly the case in relation to the parish church).

The first step in the development of the modern form can be seen in the form *Ishton* 1812–14. In spoken forms throughout south Wales *s* becomes *sh* after a long or short *i*, as in *disgwyl* > *dishgwl*, *meistr* > *mishtir*, *pris* > *prish* etc. In the English form *Easton*, the diphthong *ea* would have had virtually the same sound quality as a long W *ī* to the ears of Welsh speakers and palatalisation of *s* in common parlance under such circumstances would, and obviously did take place to produce the form *Ishton*.

Secondly, among the collected forms of the name there are several which have a prefixed *n*-: *Nishtown House* 1745, *Nishdown* 1793, *Nishton Farm* 1840, 1844–64, 1865, this being probably the remnant of the W preposition *yn* 'in' which became attached by frequent usage, prepositions being the most frequently used parts of speech with place-names. A good Welsh example is Narberth, Pembs., which is originally *Arberth*, and the process can be compared with the similar occurrence whereby the final -*n* of ME *atten* 'at the' became attached to English place-names as in the case of *Nangle* 14-17c for Angle, Nash, Norchard, Neyland, Nolton, all in Pembs., and Nash, Monknash in Glamorgan and Mon. It is significant that Eastbrook near Dinas Powys, and no great distance away from Yniston, appears in the field names *Ishbrook Meadow*, *Ishbrook Acre* in 1884, and is recorded as *Nishbrook* in 1839.

Whether the full form of the Welsh preposition *yn* played its part in the development of the form *Yniston*, *Ynyston* 1892, is not certainly known but it is possible that the form is the result of a conscious attempt to give the name a meaning, possibly by analogy with W *ynys*, as noted above.

Ynysawdre

Now the name of a civil parish, Ynysawdre was originally the name of a holding on the banks of the river Ogwr between Bridgend and Maesteg, below Abergarw and to the east of Ton-du, subsequently by the seventeenth century that of a substantial farmhouse. The present Ynysawdre Comprehensive School stands near the original site.

It is a name like many others which has presented problems to the non-Welsh scribes of official documents over the years. It is recorded as *Ynys Nawdre* 1631, *Ynys Naudre* 1692, *Ynisnawdre* 1790 etc. where the first element is W *ynys* 'island; watermeadow' which is so well-evidenced along the river banks of the main south Wales valleys as a term for land which is subject to inundation by flood-water + a form which suggests a compound of the W numeral *naw* 'nine' + *tref* 'homestead, farmstead', possibly a deliberate creation to give the name a meaning. Another proffered interpretation was that the second element was W *nawdd* 'protection, support, defence' (as in W *noddfa*) + *tref*, forming the compound *nawdd-dre(f)*, perhaps in the sense of 'sanctuary, refuge', whilst a further invention appears in the form *Ynys Awbrey* 1670.

However, another consideration arises from a comparison with the names of two farms on high ground between Pontrhydyfen and Castell-nedd (Neath), Hawdref Fawr and Hawdref Ganol (W *canol* 'middle') each, respectively, a division of an original holding, Hawdref, in the eighteenth century, so that it is very possible that it is the form *Ynys -(h)awdre(f)*, corroborated to some degree by the form *Ynis Hawdra c.* 1780 on a Dunraven estate map, which has survived in common speech as Ynysawdre.

The form *hawdref* is not related to the W adj. *hawdd* 'easy, pleasant' as there is a substantial body of evidence to suggest that its base is the W form *hafdref*, i.e. W *haf* 'summer' + *tref*, 'summer dwelling, abode', for the earlier attested forms of the names of the farms in the Neath area noted above were *Have dre* 1602, 1784, *Middle and Lower Hafdre* 1741, *Hafdre Genol* 1815 etc., *hafdre(f)* > *hawdre*, medial W *w* in compounds being liable to interchange with *f* in spoken Welsh as in *cawod/cafod*, *cywaeth/cyfoeth* etc. It is an alternative term synonymous with the common *hafod* (with *bod* 'dwelling') and the less common *hafoty* (with *tŷ* 'house') which is mainly confined to north Wales. As is well known, these are terms for locations used in the practice of transhumance in Wales, the summer dwelling as contrasted with the *hendref* (W *hen* 'old'), the 'old', original settlement or winter dwelling.

The main difficulty in accepting this interpretation is that the second element of the earlier forms of the name cited above contain an initial *n-*, as in *Ynys Nawdre* 1631. In his classic treatment of place-name evidence for the practice of transhumance in Wales, Professor Melville Richards indicated the existence of other terms which reflect the same procedure, among them a term for 'an autumn dwelling' incorporating W *cynhaeaf* 'harvest; harvest-time, autumn' + *tref*, namely *cynhaeaf-dre* > *cynhaefdre*. *Cynhaeaf* is derived from the prefix *cyn(nh)-* < Br.**kintu-* 'before' + *gaeaf* 'winter', the older form being *gaef*. This has assumed various forms as farm names and place-names, one being Cynheidre, the main element in the names of four farms and later a well-known deep colliery in the Llanelli area, Carms. Another was *Cynhowdre* (now Cefnhafdref) in Llanafan Fawr, Brecs. In Upper Lledrod, Cards., the name takes the form *(Y) Gynhawdre* (erroneously *Gwenhafdre* by the eighteenth century) and this form compounded with *ynys*, as *Ynysgynhawdre*, could conceivably survive in common parlance by loss of the unaccented third syllable as *Ynysnawdre* > *Ynysawdre*.

Ynyswilernyn

W *ynys* 'island', Ir *inis*, *inse*, the first element in this name, like its counterpart OE *ēg*, is used in a variety of senses in place-names. The most obvious is that which we today regard as normal, to use the dictionary definition, 'a piece of land surrounded by water', but, as Sir Ifor Williams has pointed out, and Margaret Gelling also in her discusssion of OE *ēg*, it can be applied to any area of ground, possibly raised, in wet or marshy surroundings. In inland situations in Welsh place-names it is used for low, flat ground or meadow alongside a river which is submerged when the river overflows its banks, sometimes termed 'a water-meadow', and often good pasture. The main river valleys of south Wales, particularly those of the Nedd and Tawe, are liberally sprinkled with compound names with *ynys* as first element, although it can also be the second element as generic in a compound.

It is in this latter sense, in all probability, that the element occurs in Ynyswilernyn, but the name itself, as far as is known, is not in common circulation nor is it particularly well-evidenced. It was the name of a farm, which no longer stands, derived from that of the comparatively narrow strip of flat meadow land (undoubtedly the *ynys* in question) on the east bank of the Tawe in the parish of Cilybebyll and opposite Godre'r Graig, north of Pontardawe. The farm is marked on the 1884 Ordnance Survey six-inch map, its location being roughly at a point where a stream falls in small cataracts down the steep slope of the side of the Tawe valley in this vicinity. High on the slope is the farm of Tarenni Gleision (W *tarenni*, a pl. form of W *tarren*, *tarran* 'hill, scarp, rocky outcrop' etc. + the pl. form of the W adj. *glas* 'blue', which has a wide range of colour senses like 'green, greyish-blue, translucent' etc. possibly referring here to the effect of a bluish haze through which a rocky outcrop could often be perceived). I have been informed on good authority that it was probably on the land of these two farms that the Tarenni Colliery (1903) was established.

The earlier recorded forms of the name have led some to postulate a personal name as its second element. They are *ynys wyliheyrnyn* 1570, *Ynys Willi hernyn* 1603, 1688, *Ynys Will hernin* 1720–1 but, fortunately, Professor Melville Richards has already proffered an interpretation which is wholly convincing in the light of two papers in which he discussed the W n.m. *chwil* 'beetle, chafer' and its derived forms as elements in Welsh place and river names.

As a simplex form it occurs in the river name Chwil, Cards. and combined with other elements in Pant-y-chwil, Carms., Pwll-y-chwil, Cards., Chwilgrug, Mon. (with W *crug* 'hillock, mound, cairn') anglicised Wilcrick. As an adj. with the ending *-og* which signifies an abundance of what is implied in the first element, Chwilog 'an abundance of beetles' which exists as a place-name in Caerns. and as a river name in Llanelli in the variant form *Chwilwg > Whilwg* (with *chw- > wh-, w-* generally in spoken Welsh in south Wales) this being also what now appears as the river name Wheelock, Cheshire. A related form, among others, may be that which appears with the ending *-ach* in Brynwhilach in the parish of Llangyfelach.

Yet another *chwil*-related form is *chwiler* 'maggot, serpent, viper' which exists as the name of a stream which flows into the Clwyd in Denbighshire at Aberchwiler (anglicised Aberwheeler). The adjectival form in *-og* of this noun is *chwileiriog*, and this exists as the name of a township in Llanarmon-yn-Iâl, Denb., and as Wileirog on the map near Bow Street, Cards. The diminutive form of *chwiler* is *chwileryn*, by the addition of the diminutive ending *-yn*, and Professor Richards was firmly of the opinion that it is a variant form of *chwileryn* which is seen as the second element of the name Ynyswilernyn. He argued that in such a form the noun *chwiler* must have acquired an excrescent *-n* after the final *-r* in common parlance, as in words like *siswrn* (ME *cisour(s)*, E *scissor(s)*), *pinsiwrn* (E *pincer(s)*), *miswrn* (ME *viso(u)r*) to give the form **chwilern*, and that it was to this derivative that the diminutive ending *-yn* was added > *chwilernyn* and *w(h)ilernyn*, another corroborative variant form *chwilernin* being evidenced in a poem by the medieval poet Dafydd ap Gwilym. The original Ynyswilernyn, therefore, may have been a portion of land characterised by an infestation of grubs or vipers.

Further Reading

B.G.Charles, *Old Norse Relations with Wales* (Cardiff, 1934).

Non-Celtic Place-Names in Wales (London, 1938).

The Place-Names of Pembrokeshire (Aberystwyth, 1992).

Michael Eyers, *The Masters of the Coalfield. People and Place-Names in Glamorgan and Gwent* (Risca 1992).

John Field, *English Field-Names. A Dictionary* (London, 1972, 1982).

M. Gelling, W.F.H. Nicolaisen, M. Richards, *The Names of Towns and Cities in Britain* (London, 1970, 1986).

M. Gelling and Ann Cole, *The Landscape of Place-Names* (Stamford, 2000).

D.M. John, *Cynon Valley Place-Names* (Llanrwst, 1998).

B.L. Jones, *Enwau* (Llanrwst,1991).

Yn ei Elfen (Llanrwst, 1992).

G.T. Jones, Tomos Roberts, *The Place-Names of Anglesey* (Isle of Anglesey County Council, 1996).

Henry Loyn, *The Vikings in Wales* (London, 1976).

R. Morgan, *A Study of Radnorshire Place-Names* (Llanrwst, 1998).

with R.F.Peter Powell, *A Study of Breconshire Place-Names* (Llanrwst, 1999).

H.W. Owen, *The Place-Names of East Flintshire* (Cardiff, 1994).

A Pocket Guide. The Place-Names of Wales (Cardiff, 1998).

O.J. Padel, *A Popular Dictionary of Cornish Place-Names* (Penzance, 1988).

G.O. Pierce, 'Place-Names', in J.F.Rees (ed.), *The Cardiff Region. A Survey* (Cardiff, 1960), 171–6.

The Place-Names of Dinas Powys Hundred (Cardiff, 1968).

'Mynegbyst i'r Gorffennol' in Ieuan M.Williams (ed.), *Abertawe a'r Cylch* (Llandybie, 1982), 73–98.

'The Evidence of Place-Names' in H.N.Savory (ed.), *The Glamorgan County History*, Vol. II, Appendix 2 (Cardiff, 1984), 456–92.

'Place-Names' in D.H.Owen (ed.), *Settlement and Society in Wales* (Cardiff, 1989), 73–94.

M. Richards, 'Norse Place-Names in Wales' in *Proceedings of the International Congress of Celtic Studies, Dublin, 1959* (1962), 51–60.

Welsh Administrative and Territorial Units (Cardiff, 1969). Essential for the purposes of identification and correct spelling of Welsh place-names.

Enwau Tir a Gwlad (Caernarfon, 1998).

Adrian Room, *A Concise Dictionary of Modern Place-Names in Great Britain and Ireland* (Oxford, 1983).

R.J. Thomas, *Enwau Afonydd a Nentydd Cymru* (Caerdydd, 1938).

Ifor Williams, *Enwau Lleoedd* (Lerpwl, 1945, 1969).

A detailed bibliography of publications on place-names in Wales is contained in Jeffrey Spittal and John Field, *A Reader's Guide to the Place-Names of the United Kingdom* (Stamford, 1990).

Index

Aber-annell 204
Aberbaiden 11-12
Abercegin (Caerns.) 90
Abercegyr (Mont.) 91
Aberchwiler (Denbs.) 225
Abercraf (Brecs.) 21
Abercynffig 63
Aberdâr: *see* Aberdare
Aberdare 15, 21, 132, 150-1, 199
   Lord: *see* Bruce, H.A.
Aberffrwd 15
Abergeirw (Merioneth) 135
Abergwaun (Pembs.) 219
Abergwrelych 12-13
Abernant Ironworks 215
Abernant y Wenallt 215
Aberpennar 13-16
Abertawe: *see* Swansea
Aberthaw 12, 110
Abertysswg 161
Aberwheeler (Denbs.) 225
Afon Genllysg 131
Ala, Yr (Caerns.) 61
Ala Las (Caerns.) 61
Alafowlia (Denbs.) 61
Album Monasterium 218
Amlwch (Anglesey) 13
Amroth (Pembs.) 13, 173-4
Angelton 202-4
Angle (Pembs.) 222
Anglesey 178, 182
Arannell 204
Argae 135
Arthur, King 11, 16-17, 90-1
Arthur's Butts Hill 16-17
Arthur's Stone 16-17
Atkin, Mary 98

Bachygwreiddyn 171
Bachynbyd 126
Baglan 103, 117
Balden, Nicholas 50
Banadlog 86
Banalog 86
Banff (Scotland) 12
Bangor (Caerns.) 90
Bannau Brycheiniog 207
Bardsey 178, 182
Bargoed 21, 53
Barnstaple (Devon) 61

Barry 47, 56, 186
Basset family 18
Bathafarn 126
Bawdrem/Bawdrip, Thomas 180
   William 179-80
Beacon Hill 207
Beaufort (Mon.) 162
Beaupre 18-19
Beaupre (Hunts.) 19
Beaurepaire (Hants.) 19
Beddgelert (Caerns.) 77
Bedlinog 19, 214
Bedwas (Mon.) 89
Bedwellte: *see* Bedwellty
Bedwellty (Mon.) 15, 19-20, 214
Bedwenarth 19, 214
Bellasallagh (Ireland) 56
Bellimoor (Herefs.) 22
Belper (Derbys.) 19
Benllech, Y (Anglesey) 114
Berthgelyn 148
Bethesda (Caerns.) 131
Beting 36-7
Bettws 161
Betws 161
Betws Garmon (Caerns.) 20
Bevan, John 196
Bewper (Kent) 19
Bewper (Surrey) 19
Bewpyr, Y 18-19
Birch, W. de Gray 73
Bishopston 183
Blaen-annell 204
Blaenbradach 193
Blaencynffig 63
Blaenegel 20-1
Blaenllechau 142-3
Blaenyrangell (Pembs.) 203
Blaenyrolchfa Fach/Fawr 134
Bochrwd (Brecs.) 39
Bodedern (Anglesey) 19
Bodewryd (Anglesey) 19
Bodffordd (Anglesey) 213-15
Bodorgan (Anglesey) 19
Bodringallt 19, 214
Bolgoed 21-2
Bonvilston 195, 198
Boughrood (Brecs.) 39
Bow Street (Cards.) 171, 225
Bowdler, Thomas 164

Bowen, Emanuel 16, 23, 52, 78, 84, 118
Bradney, Sir Joseph 62, 123, 147
Brecon Beacons 207
Bridgend 103, 202
Bristol, St Augustine's abbey in 59, 62, 192
Briton Ferry 81
Brittanny 27-8, 111
Brogynin (Cards.) 79
Brombil, Y 22-3
Broth Lane 217
Bruce, H.A. 1st Lord Aberdare 14
Brychan Brycheiniog 20
Bryn Banadl 23
Bryn Banal 23
Bryn Byrbais 51
Bryn Dadlau 24
Bryn Llechwenddiddos 115
Bryn Llwynaeddan 118
Bryn-llecharth 33
Bryn-y-don 144
Brynbolgoed 22
Bryncethin 215
Bryndown 49, 144
Bryngurnos 88
Bryngyrnos 89
Brynhill 49, 144
Brynmawr (Mon.) 88
Brynwhilach 225
Brynygynnen 23-4
Buarth (Caerns.) 32
Bugeildy (Rads.) 207
Burford (Oxon.) 58
Burgate (Suffolk) 58
Bwlch y Ddeufaen (Caerns.) 67

Cadoxton 186
Cadoxton-juxta-Neath 165
Cadrawd: *see* Evans, T.C.
Cae'rarfau 30-1
Caerau 63-4, 200
Caerdydd: *see* Cardiff
Caerffili: *see* Caerphilly
Caerleon (Mon.) 73
Caernarfon (Caerns.) 61
Caerphilly 19, 27-30
Caesar's Arms 131
Caesar's River 131-2
Caeyrarfau 30-1
Caldy 182
Cameron, Kenneth 100
Candleston 195, 197-8
Canton 140
Capel Fynachlog 73
Capel Illtud (Brecs.) 108
Capel Llanilltern 96
Capel y Fanhalog 73
Cardiff 25-7, 53, 94, 97-8, 99, 139-41, 171, 178, 179-80, 192-5, 216-17
Cardiff Little Heath 42
Carmarthen, Black Book of 38, 155
Carn Llechart 32-3
Carnau Cegin (Carms.) 90
Carnedd Arthur (Caerns.) 16
Carnedd y Filiast 185
Carneddi Llwydion 32
Carnllwyd 44
Cas-fuwch (Pembs.) 77
Casllwchwr 33-4
Castan 21
Castell Blaidd 35
Castell Corryn 35
Castell Crychydd 35
Castell Hendre (Pembs.) 83
Castell y mwnws 35
Castell-y-bwch (Mon.) 35-6
Castell-y-dryw 35
Castell-y-geifr (Cards.) 36
Castellau 74
Castlebythe (Pembs.) 77
Castlemartin (Pembs.) 177
Cefn Betingau 36-7
Cefn Bryn 17
Cefn-coed 63, 83
Cefn-faes 83
Cefn-rhos 83
Cefncelfi 37-8
Cefneithin (Carms.) 31
Cefnhafdref (Brecs.) 224
Cefnllwynau 188
Cefnmabli 54
Cefnmachen 124
Cefnpennar 14
Cefnsaeson 214
Cefnydfa 38-9
Cefn(y)don 63
Ceidio 103
Celydd Ifan 40
Celynen 21
Cemais (Mont.) 13
Cencoed 63, 188
Cendl, Y (Mon.) 162
Cerdin 21
Cerrig Arthur (Caerns.) 16
Cerrig Cegin (Carms.) 90
Charles, B.G. 45, 51, 83, 89, 95, 105, 113, 174, 177, 220
Cheriton 186
Childs, Jeff 82
Chwil (Cards.) 225
Chwilgrug (Mon.) 225
Chwilog (Caerns./Carms.) 13, 225
Cil-fai 126
Cilgeti (Pembs.) 161
Cilibion 40-2, 165
Cilybebyll 37, 40, 81-2, 139, 224

Cimla 42-3
Clare, Gilbert de 27
Cleirwy (Radnor) 77
Clenennau (Caerns.) 187-8
Clun(y)dadlau 24
Clydach 104
Clynnog 86
Clynnog Fawr/Fechan (Anglesey/Caerns.) 187
Clyro: see Cleirwy
Cnap Coch 48
Cnap Llwyd 48
Cnicht, Y 84
Cnydfa 39
Cocket 50
Coed Craig Ruperra 175
Coed Cynllan 188
Coed Ffos-ceibr 149
Coed Horseland 36
Coed Melwas 121
Coed Raglan 44
Coed Riglan 74
Coed Twyncyn 208
Coed y Capel 209
Coed y Cradock 51
Coed y Cymdda 42-3
Coed y cynllwyn 188
Coed y Devonald (Pembs.) 51
Coed-y-gaer 104
Coedarhydyglyn 43-4
Coedriglan 43-4, 74
Coedygores 44-5
Coedymwstwr 45-7, 104
Coedypebyll 104
Coeten Arthur 16
Coety: see Coity
Cogan 62
Coity 45, 103-4, 121
Cold Knap 47-8
Coldra (Mon.) 179
Corndon Hill (Mont.) 50
Corntown 49-50
Corringdon (Devon) 49
Cortwn 49-50
Corwen (Merioneth) 47, 53, 114
Courtyrala 60
Cowbridge 195
Coxe, William 62, 148
Coychurch 45, 74, 139
Crafnant 21
Craig Ruperra 175
Craig-cefn-parc 97, 104
Craigybwldan 50-1
Crediton (Devon) 195
Creigiau 30, 131
Creunant Forest 67
Cribarth 32
Croes Cwrlwys 51-3

Croescadarn 53-5
Crofft-yr-haidd 196
Crofftygenau 55-6
Crosskeys (Mon.) 15
Crwys, Y 97
Crwys Bychan, Y 97
Crwys Mawr, Y 98
Crwys Road 97-8
Crynfai 126
Culverhouse Cross 51-3, 64
Cwm Cynffig 63
Cwm Sart 191
Cwm Talwg 56-7
Cwm y Brombil 204
Cwm-hir Abbey (Radnor) 77
Cwmbwrla 57-8
Cwmcidi 56-7
Cwmergyr (Cards.) 189
Cwmwbwb 24, 199
Cwrt Sart 59, 165
Cwrt-y-fil 58-60, 212
Cwrtrhydhir 59
Cwrtybetws 59
Cwrtycarnau 59
Cwrtyrala 60-1
Cymdda Mawr 43
Cymer Abbey (Merioneth) 150
Cymin, Y 61-2
Cyncoed 63, 83, 188
Cynffig 62-3
Cynheidre (Carms.) 223-4
Cynllwyn 188
Cynllwyn-du 188
Cynllwynau 188
Cyntwell 63-5
Cynwyl Elfed (Carms.) 67
Cyrniau, Y 88, 142

Dafydd ap Gwilym 121, 194, 225
Dafydd Benwyn 37, 96, 180
Dafydd Morganwg: see Morganwg, Dafydd
Danycapel 124
Dâr 21
Davies, A.T. 66
    J.B. 96, 167
    J.D. 185
    John 88
    Wendy 76
Debateable Land (Cumb.) 199
Dinas Mawddwy (Merioneth) 37
Dinas Powys 57, 177-9, 207
Disgwylfa 123, 126
Dodridge (Devon) 195
Dolwyddelan (Caerns.) 14, 67
Dowlais 57
Droop (Dorset) 67
Drope 66-7, 178
Drupe (Devon) 67

Drysgoed 68
Drysgol 67-9
Drysgol Goch, Y (Carms.) 67
Dryslwyn (Carms.) 68, 143
Duffryn Aberdare 14-15
Dulais 57
Dunkery Beacon (Som.) 207
Dyffryn Aberdare 14-15
Dyrysgol (Merioneth) 68

Eastbrook 141, 222
Easton 141
Ebbw Vale (Mon.) 162
Eglwys Newydd, Yr 217-19
Eglwysgeinwyr 124
Eglwysilan 32, 41, 214
Eifionydd 29, 122
Elbred, Rachel 65
Ellis, D. Machreth 135
Ergyd Isaf/Uchaf 189
Ergyd Non (Cards.) 190
Ergyd Ronw (Merioneth) 190
Ergyd-y-gwynt 189
Esgair Annell 204
Etherington, John 86
Evans, David 84
  T.C. 62
Ewenni 70-1, 210

Fagwyr, Y 104
Fanhadlog, Y 86
Fanhalog, Y 72-4
Fanhaulog 72
Farmers (Carms.) 15
Felenrhyd, Y (Merioneth) 167
Ferndale 143
Fernhill Farm 213
Fetters (Fife) 138
Ffald Farm 130
Ffald-y-dre 154
Fforchegel 20
Ffordd y brain 145
Fforest-fach 118
Ffrwdwyllt 204
Ffynnon Dwym 131
Ffynnon Gegin Arthur (Caerns.) 16, 90
Ffynnon Ofer 128
Ffynnon y Fynachlog 74
Fishguard (Pembs.) 219
Fishlake 58
Flemingston 195, 200
Fleur-de-Lis (Mon.) 15
Foithear (Inverness) 138
Fonmon 49, 95
Fontygary 49
Foreland 212
Forlan, Y 130
Forty 74-5, 212

Foyers (Inverness) 138
Fulford 56, 166
Fullabrook 56
Fullaford 56
Fullamoor 56

Gabalfa 76-7
Galon Uchaf 78-9
Gamage, Robert 121
Garnhill 144
Garth Graban 33
Garth Hill 16-17
Garth Maelwg 33
Gelli-hir 41
Gelliargwellt 79-80
Gelligaer 31, 42, 80-2, 87, 153, 188
Gellihirion 41
Gelling, Margaret 58, 84, 144, 157, 214, 219, 224
Gelli'r Gnydfa (Merioneth) 39
Gelli'rhaidd 196
Gendros 63, 82-3
Geoffrey of Monmouth 121
Gernos (Pembs.) 89
Gernos Mountain (Cards.) 88
Gilbert, The 156
Gileston 49
Glais, Y 57
Glanaman (Carms.) 67
Glanffrwd: *see* Thomas, William
Glyn Rhedynog 143
Glyn Tarell (Brecs.) 22, 89
Glynneath 12
Glynnog, (Y) 187
Glynrhigos 89
Glywysing 29
Gnol, Y 84-5
Gnydfa, Y 39
Gofilon (Mon.) 12
Golchfa 'Ralltlwyd (Merioneth) 135
Gorfynydd 29
Gors, Y 86, 90
Gorwenydd 29
Gosbeck 200
Gosbrook 200
Gosford 200
Gosforth 200
Goston 200-1
Gostrode 200
Goswick (Northumberland) 200
Govilon (Mon.) 92
Gowerton 88
Gowlog 85-6
Gray-Jones, Arthur 162
Great Orme's Head (Caerns.) 219
Greenyard 156
Grennett, John 156
Groes, Y 23, 86

Groes-faen 86-7
Groneath 29
Gronedd 29
Gruffudd ap Cynan 219
Gruffudd Hiraethog 19
Gruffydd, Geraint 79
  W.J. 29
Gurn Ddu, Y (Caerns.) 88, 142
Gurn Goch, Y (Caerns.) 88, 142
Gurn Las, Y (Caerns.) 88, 142
Gurnos, Y 88-9
Gwal y Filiast 185
Gwedir (Caerns.) 138
Gwendraeth Fach (Carms.) 163
Gwenfo 175
Gwern yr Havod Talog 94
Gwernga 130
Gwernycegin 89-91
Gwernyclepa (Mon.) 194
Gwreiddyn 170
Gwydir (Caerns.) 138
Gwydr Crescent 137
Gwydr Mews 137
Gwydr Square 137
Gwynfa 126
Gwynfai 126
Gwynionydd 29
Gwynlais 57
Gwynllwg 29
Gwyr y Mera 127
Gwytherin (Denbs.) 103
Gyfeillion, Y 91-2
Gyfynys (Caerns.) 92
Gyrn, Y (Brecs.) 89, 142
Gyrn (Denbs.) 142
Gyrnos (Brecs.) 88

Hackerford 170
Hafod Deca 94
Hafod Decaf, Yr 94
Hafod Decca 94
Hafod Ruffudd 95
Hafodhalog 93-4
Hafodwgan 94-5
Hall, Benjamin Lord Llanofer 128
Harlech (Merioneth) 114
Harris, B.D. 125
  Howell 120, 201, 213
Harry, Richard 96
Hatford (Berks.) 144
Haverfordwest (Pembs.) 215
Havod Talog, Yr (Merioneth) 94
Hawdref Fawr/Ganol 223
Heath, Little 42
Hen Lodre Eliddon 117
Hendre Forgan 95
Henllys (Mon.) 31
Henry's Moat (Pembs.) 83

Henstaff 96
Heol Llechwenddiddos 115
Heol-lace 150-1
Heol-las 150-1
Heol-y-cawl 216-17
Heol-y-crwys 97-8
Heol-y-march 98-9, 210
Hereford 215
Highlight 209-11
Hirwaun 21-2, 216
Hirwaun Bolgoed 22
Holland Arms (Anglesey) 15
Holton le Clay (Lincs.) 107
Homri 99-100
Hopkin, Lewis 115
Hornby (Lancs./Yorks.) 99
Horseland Farm 35-6
Hospitallers, Knights 64-5
Hyssington (Mont.) 50

Ifor ap Llywelyn 194
Ilston 108, 183
Iolo Goch 191
Iolo Morganwg: *see* Morganwg, Iolo

Jackson, Kenneth 70, 164
James, Brian 221
  Edward 31
Jenkins, R.T. 78
John, Deric 158
Jones, Bedwyr Lewis 20, 37
  David 201
  D.W. 11, 13, 58, 73, 78, 130
  Francis 176
  Griffith 11, 23, 201, 202
  John 14
  Thomas 38, 40

Kendall, Edward 162
  Jonathan 162
Kenfig 58, 62-3
Kilgetty (Pembs.) 161
Kingcoed 83
Knights Hospitallers 64-5
Knoll, The 84-5
Kymin, The 61-2

Laleston 200
Lamby 99
Lampeter (Cards.) 158
Lampha 104-5
Lampha (Pembs.) 105
Lamphey (Pembs.) 105
Landevennec (Brittany) 111
Larnog 101-2
Latts 127
Lavernock 101-2
Laverstock (Som.) 101

Laverstoke (Som.) 101
Laverton (Som.) 101
Leckwith 43, 102-3, 141, 212, 216
Lecwydd: see Leckwith
Leland, John 14, 22, 24, 56, 63, 122, 157, 167, 178, 180, 218
Lewis, Edward 180
  Henry 104
  Rice 24, 180
  Samuel 58, 102, 176, 200
  Saunders 102
Lhuyd, Edward 94, 148, 150, 207
Lisvane 87, 151
Lisworney 29, 121-2
Litchard 103-4
Little Nash 163
Llafar 168
Llan-crwys (Carms.) 155
Llan-gain (Carms.) 124
Llan-gan 200-1
Llan-gors lake (Brecs.) 127-8
Llan-non (Carms.) 23
Llanafan Fawr (Brecs.) 88, 224
Llanarmon-yn-Ial (Denbs.) 225
Llanbedr Pont Steffan (Cards.) 158
Llanbethery 110
Llanblethian 110
Llanboidy (Carms.) 116
Llanbradach 193-4
Llanbradach Fawr 114
Llancadle 75
Llancaeach 193
Llancarfan 51, 75, 85-6, 98, 109, 198
Llandaff 23-4, 26, 105
  bishop of: see Ollivant, Alfred
  Bishop's Court in 24
  Castle in 24
  Cathedral School in 23, 24
  Llandaff Court in 24
  Llys Esgob in 24
  Old Bishop's Palace in 24
Llandaff North 218
Llanddeiniolen (Caerns.) 90
Llanddewi 50
Llanddewi'r Crwys (Carms.) 97
Llandecwyn (Merioneth) 94
Llandegái (Caerns.) 105
Llandeilo 109
Llandeilo Fawr (Carms.) 88, 90, 105
Llandeilo Tal-y-bont 21-2, 85
Llandenny (Mon.) 47
Llandinam (Mont.) 39, 207
Llandoch(au) 34
Llandough 34
Llandow 42
Llandre (Cards.) 117
Llandremor 117
Llandudno (Caerns.) 109
Llandudwg 106
Llandwrog (Caerns.) 109
Llandyfái 105
Llandyfeisant (Carms.) 105
Llandyfodwg 115
Llandygai (Caerns.) 37, 105
Llandyssul 161
Llandysul 161
Llanedern 44, 53, 77
Llanedi (Carms.) 95
Llanegryn (Merioneth) 39
Llanegwad (Carms.) 125
Llanelli (Carms.) 39
Llanelltud (Merioneth) 107
Llanenewyr 110-11
Llanfabon 15, 37, 114
Llanfairfechan (Caerns.) 67
Llanffa 104-5, 105
Llanfihangel Abergwesyn (Brecs.) 31
Llanfihangel ar Elái 106
Llanfihangel y fedw 106
Llanfihangel Rhos-y-corn (Carms.) 77
Llanfihangel-y-pwll 105-7
Llanfoy (Herefs.) 105
Llanfyfor 128-9
Llanfyrnach (Pembs.) 89
Llangathen (Carms.) 143
Llangeinor 117, 124, 166
Llangeinwyry 124
Llangeler (Carms.) 39, 125
Llangennech (Carms.) 117
Llangennith: see Llangynydd
Llangiwg 20, 36-7
Llangoed (Anglesey) 37
Llangrallo 45-6, 139
Llangwarran (Pembs.) 120
Llangwathen (Pembs.) 120
Llangyfelach 36, 48, 155, 225
Llangyndeyrn (Carms.) 163
Llangynidr (Brecs.) 77
Llangynllo (Cards.) 88
Llangynnwr (Carms.) 110
Llangynog (Carms.) 37
Llangynwyd 11, 56
Llangynydd 30
Llanharan 125, 126, 188
Llanhari 188
Llanhywel (Rads.) 120
Llanilltud (Brecs.) 108
Llanilltud Fach 108
Llanilltud Faerdre(f) 51, 68, 108, 185, 217
Llanilltud Fawr 44, 107-8
Llanilltud Gwyr 108
Llanilltud Nedd 108
Llanllechid (Caerns.) 67
Llanllwchaearn (Cards.) 77
Llanmaes 87
Llannerchyryrfa 31

Llanover (Mon.) 128
Llanrhaeadr yng Ngheinmeirch (Denbs.) 90
Llanrhaeadr-ym-Mochnant (Denbs.) 189
Llanrheithan (Pembs.) 98-9
Llanrhidian 64, 86, 94, 110
Llanrumney 193-4
Llanrwst (Denbs.) 128
Llansamlet 48
Llansanffraid (Cards.) 190
Llansawel 81
Llantarnam (Mon.) 73, 109
Llanthony 109
Llantood (Pembs.) 108
Llantriddyd: *see* Llantrithyd
Llantrisant 35, 42, 51, 87, 120-1, 139, 166, 187, 191, 196, 215
Llantrithyd 12, 35, 108-10
Llantwd (Pembs.) 108
Llantwit Fardre 68, 108
Llantwit juxta Neath 83, 108
Llantwit Major 44, 107-8
Llanuwchllyn (Merioneth) 150
Llanvetherine (Mon.) 188
Llanwnda (Pembs.) 110
Llanwnnwr (Pembs.) 110
Llanwrin (Mont.) 122, 128
Llanwynno 72-4, 114, 142, 148, 166, 170, 187
Llanycrwys (Carms.) 97
Llanynewyr 64
Llanyrnewydd 110-11
Llawennant 112-13
Llebenydd 29
Llech y Filiast 185
Llecharth 33
Llechwen 114-15
Llechwen, Y 114
Llechwenlydan 114
Lledglawdd 116
Lledrod, Upper (Cards.) 224
Llefyn, Y (Caerns.) 67
Llety Brongu 117
Llety'r Filiast 185
Llidiardau 104
Llodre 117-18
Llodrog 64
Llotrog 117
Llowes: *see* LLywes
Llugwy, river 34
Llwydiarth (Anglesey) 32
Llwyn Bedw 118
Llwyn Celyn 118
Llwyn Derw 118
Llwyn Eithin 118, 118-19
Llwyn Helyg 118
Llwyn-bedw 147
Llwyn-helyg 147
Llwyn-onn 147

Llwyncelyn 187
Llwyncynhwyra 124
Llwyncynllwyn 188
Llwyndyrys 68, 143
Llwyneliddon 110, 119-20, 198
Llwyngoras 45
Llwyngwarran (Pembs.) 120
Llwyngwathen (Pembs.) 120
Llwynhywel (Rads.) 120
Llwynmilwas 120-1
Llwynydadlau 199
Llwynygorras 45
Llyn Fawr 12
Llyn yr Olchfa (Merioneth) 135
Llynfi Valley 117
Llys-faen 87, 151
Llysfaen (Denbs.) 53
Llyswyrny 121-2
Llywernog (Cards.) 101
Llywes (Radnor) 103
Login 94
Logyn 94
Loughor 33-4, 63
Lower Race 162-3
Lugar, river 34
Lugg, river 34

Machen (Mon.) 123-4, 126, 215
Machynlleth (Mont.) 77, 123, 126, 215
Mackworth, Sir Humphrey 84, 127
Madley (Herefs.) 22
Maelienydd 29
Maen Ceti 17
Maenclochog (Pembs.) 87
Maenordeilo (Carms.) 138
Maentwrog (Merioneth) 87, 167
Maes Siward 98, 125
Maes-y-ward 124-5
Maes-yr-yrfa (Carms.) 31
Maeslon 144
Maesywerfa 215
Malkin, B.H. 44, 220
Mallwyd (Mont.) 123
Mansel, Philip 50
   Sir Rice 181
Margam Abbey 73, 93, 94-5, 97, 170, 181, 202, 204
Marlais 57
Martletwy (Pembs.) 105
Mathafarn (Mont.) 123, 215
Mathew, Sir David 24
   Jennet 96
   Miles 24
   Thomas 24
   family, of Radyr 24
Mathrafal (Mont.) 123
Mechain (Mont.) 123
Meerbrook (Staffs.) 113

Meersbrook (Derbys.) 113
Mefenydd 29, 122
Meiarth 126
Meidrum (Carms.) 126
Meifod (Mont.) 126
Meirionnydd 29, 122
Meiros 125-6
Meline (Pembs.) 89
Mellor (Ches./Lancs.) 154
Mera, Y 127-8
Merched y Mera 127
Merrick, Rice 24, 33, 39, 44, 84-5, 102, 134, 155, 158, 174, 193, 220, 221
Merry Brook (Worcs.) 113
Merthyr Dyfan 56
Merthyr Mawr 128-9, 170
Merthyr Myfor 129
Merthyr Tydfil 78-9, 88, 130
Mertyn 49
Michaelston le Pit 60, 105-7
Michaelston super Ely 44, 65, 106
Michaelston y fedw 106
Milton 65
Monknash 222
Monmouth (Mon.) 62
Morgan, Prys 54
   T.J. 54
Morgannwg 29
Morganwg, Dafydd 11, 108
   Iolo 40, 48, 58, 59, 60, 86, 108, 188, 200
Morlais 57
Morlanga 129-30
Morriston 86
Mortlake 58
Mostyn (Flints.) 49
Mount, The 57
Mount Badon 11
Mountain Ash 13-16
Mulfra (Cornwall) 154
Mulvra (Cornwall) 154
Myddfai (Carns.) 126
Mynachdy 73
Mynydd Badon 11
Mynydd Carnllechart 32
Mynydd Margam 63
Mynydd Meiros 125
Mynydd Pen-cyrn (Brecs.) 142
Mynydd-y-Garth 33
Mynyddislwyn (Mon.) 14, 146-8

Nant Boeth, Y 131
Nant Bran 150-1
Nant Brane 150-1
Nant Fawr 168-9
Nant Iaen 131
Nant Iorwerth Goch 11
Nant Oer 131

Nant Talwg 56, 57
Nant Tredodridge 196
Nant Twymyn (Mont.) 131
Nant Tynyplancau 149
Nant y Cesair 131-2
Nant y Sgwrfa 162
Nant yr eira 131
Nantyffyllon 132-3
Narberth (Pembs.) 222
Nash (Glam.) 222
Nash Manor 163
Nash (Pembs.) 222
Neath 42, 63, 84-5, 89, 127-8
   Abbey 41, 59, 165
   river 12
Nefydd: *see* Roberts, William
Nelson 15
Nennius 11
Nevern (Pembs.) 89
Newbridge (Mon.) 14
Newcastle (Glam.) 202
Newton le Willows 107
Neyland (Pembs.) 222
No Mans Land (Wilts.) 199
Nolton (Pembs.) 222
Norchard (Pembs.) 222
Norton 49
Nurston 198

Odnant 131
Odyn's Fee 180
Oerddwr 131
Oernant 131
Ogof y Pebyll 139
Olchfa, Yr 134-5
Old Beaupre 18-19
Ollivant, Alfred bishop of Llandaff 24
Olmarch Cefn-y-coed (Cards.) 99
Olmarch Fawr (Cards.) 99
Olmarch Ganol (Cards.) 99
Olmarch Isaf (Cards.) 99
Olmarch (Pembs.) 98-9
Otterford (Som.) 17
Owen, Hywel W. 165
Owen-Pughe, William 60, 195
Oxford 172
Oystermouth 37, 165, 184, 186

Padel, Oliver 28
Pant y Philip (Pembs.) 51
Pant y Rhas 163
Pant-y-chwil (Carms.) 225
Pant-y-crwys 97
Pant-y-gwydr Road 137
Pantaquesta 136-7
Pantygwydir 137-8
Pantygwydir Road 137
Pantynenbren 148

Pantyryrfa (Mon.) 31
Parc Coed Marchan 96
Parc-y-Prat (Pembs.) 51
Pebyll 138-9
Pebyll-y-brig 139
Peebles 139
Pegwn Bach/Mawr 207
Pembrey (Carms.) 13
Pen Llwyn Aeddan 118
Pen y Gurnos (Cards.) 88-9
Pen-bol (Anglesey) 22
Pen-bre (Carms.) 117
Pen-coed 45, 65, 104, 201
Pen-cyrn 88, 141-2
Pen-llin 198
Pen-rhas 163
Pen-y-fai 126, 202
Pen-y-groes 86
Pen-y-gyrn (Mont.) 142
Pen-yr-ala (Denbs.) 61
Pen-yr-ergyd 189
Penalltau 91
Penarth 33, 58-9, 61-2
Pencisely 139-41
Penclawdd 97, 117
Pencoed 45, 65, 104, 201
Pendeulwyn 21-2, 39, 149
Pendown 145
Pendoylan 21-2, 39, 149
Pendyrys 142-3
Penegoes (Mont.) 77
Penhefyd 144-5
Penhill 49, 140
Penlle'rbebyll 139, 146
Penlle'rbrain 145-6
Penlle'rcastell 146
Penlle'rgaer 146
Penlle'rneuadd 146
Penllwyn (Mon.) 146-7
Penllwyn-sarff (Mon.) 146-8
Penllwynaeddan 118
Penllwynbrain 146
Penmachno (Caerns.) 14
Penmark 95, 163, 180, 198
Penrhiw-ceibr 148-9
Penrhiw'rceibr 149, 175
Penrhiw'rgwiail 149, 175
Penrhiwtyn 149-50
Penrhyn Gwyr 220
Penrhys 73
Penrikyber 149
Pentre Eiriannell (Anglesey) 203
Pentrebane 150-1
Pentrehaearn 151-2
Pentyle'rbrain 146
Pentyrch 16, 30, 87, 120-1, 131-2, 187
Penypyrod 220
Peterston super Ely 96, 129-30

Peulwys (Denbs.) 53
Philleigh (Cornwall) 28
Phillimore, Egerton 111, 142
Phillips, D.R. 127
Pindrup (Gloucs.) 67
Plas Cilybebyll 81
Plas Coedymwstwr 46
Plas Milfre 153-4
Plas Watford 213-14
Plounéour-Lanvern (Finistère) 111
Plounéour-Menez (Finistère) 111
Plounéour-Trez (Finistère) 111
Plwcahalog 94
Plymouth estate 59
Pont Bwrlac 58
Pont Llewitha 154-5
Pont Walby 12
Pont y Rhidyll 156
Pont-rhyd-y-fen 157-8
Pont-y-gwaith 143
Pontardawe 158
Pontarddulais 117, 158
Pontardulas 158
Ponterwyd (Cards.) 101, 189
Pontprennau 169
Pontyberem (Carms.) 23
Pontypool (Mon.) 162-3
Porridge Lane 217
Port Penrhyn (Caerns.) 90
Port Talbot 23
Porthcawl 217
Porthkerry 163
Portmead 113
Pottage Lane 217
Powell, Evan 19
Prestatyn (Flints.) 49
Price, Thomas 213
Pughe: see Owen-Pughe
Pulford (Cheshire) 215
Pwll Cam 97
Pwll Rhaslas 163
Pwll-y-chwil (Cards.) 225
Pwllheli 61
Pwllygwreiddyn 170

Race Farm 163
Radur: see Radyr
Radyr 24, 68-9, 159-60
Radyr Chain 160-1
Raglan, Sir John 44
Rale(i)gh family 53, 61
Rammell, T.W. 15
Ramsey (Pembs.) 178, 182
Rasau 161-3
Rassau 161
Raven Hill 145
Rawley family 61
Reddings Farm (Mon.) 165

Rees, William 27
Regal 20
Resolven 42
Reynoldston 37
Rhanallt Street 23, 204
Rhas Cottage 163
Rhas Fach 163
Rhaslas 163
Rhath, Y 173
Rheola 210-11
Rhigos 12, 89
Rhiw y werfa 215-16
Rhiw-wden 150
Rhiwabon (Denbs.) 175
Rhiwau 175
Rhiwbina 182
Rhiwderyn (Mon.) 175
Rhiwfallen 175
Rhiwgriafol 175
Rhiwonnen 149, 175
Rhiw'r-hwch 186
Rhiw'rperrai 149, 175
Rhodfa'r Brain 145
Rhoose 163-4
Rhos Is Dulas 164
Rhos Ymryson 24
Rhos-y-bol (Anglesey) 22
Rhosili 213
Rhosygwidir (Pembs.) 138
Rhosyryrfa 31
Rhuthun (Denbs.) 37
Rhws, Y 163-4
Rhyd Wilym 95
Rhyd-y-felin 166
Rhydding 165
Rhyddings, The 164-5
Rhyddings Park Road 164
Rhyddings Terrace 164
Rhydfelen 165-7
Rhydhalog 94
Rhydlafar 167-8
Rhydybilwg 168-70
Rhydychen 172
Rhyd(y)felin 165-6
Rhydygwreiddyn 170-1
Rhydypennau 171-2
Rhydyregel 20
Rhymney (Mon.) 15, 192-5
  Ironworks 15
Rhymni 192-5
Rhyndwyglydach 32, 128, 190
Rhytalog 94
Richards, Melville 13, 14, 20, 29, 39, 53,
  95, 122, 125, 126, 169, 219, 223, 225
  Thomas 87, 169
Richardson, J.C. 138
Rissland Farm 35, 36
Rivet, A.L.F. 173

Roath 172-4
Roberts, Tomos 22, 91, 126, 214
  William 166
Robin Hood 17
Robin Hood's Butts (Herefs.) 17
Roos (Yorks.) 164
Roose (Lancs.) 164
Roose (Pembs.) 164
Roseland (Cornwall) 28
Ross (Herefs.) 164
Rowlston (Herefs.) 148
Royton (Lancs.) 197
Ruabon (Denbs.) 175
Rug, Y 86
Rugos 12
Rumney (Mon.) 192-5
Ruperra 174-5
Rushland Farm 35, 36
Ruthun (Denbs.) 37
Ruyton (Shropshire) 197
Ryton 196-7

St Andrews 135, 141, 212, 217
St Brides 49
St Brides Major 43
St Brides Minor 66
St Brides super Ely 66, 96, 212
St Cenydd 27-30
St David's (Pembs.) 13
St Dogmaels (Pembs.) 89
St Fagans 44, 51, 52, 144, 151
St Georges 43, 44, 66, 178
St Govor's Well 128
St Hilary 18, 49
St John, Knights Hospitallers of the Order
  of 64-5
St Lythans 43, 119-20, 185, 198
St Lythan's Down 204-6
St Nicholas 44, 99, 170, 208
Sanctuary, The 64-5
Sarn Helen 210
Sblot, Y 179-80
Schwyll 176-7
Scouring Brook 162
Senghennydd 27-30
Serwynydd 29
Sgeti 161
Shee Well, The 176
Shew Well, The 176
Shiplake 58
Sketty 17, 77, 117, 134, 161
Slebech (Pembs.) 64
Smith, Colin 173
Soudrey 177-9
Southra 177-9
Southrow (Pembs.) 177
Splot 179-80
Spurgeon, C.J. 35

Stormy Down 180-1
Stradling, Edward 79-80
  family 165
  Sir John 126
Strata Florida (Cards.) 112
Strata Marcella (Mont.) 95, 112
Stratford le Bow 107
Sturmi, Geoffrey 181
Swale Bank (Sussex) 177
Swale river 177
Swansea 57, 77, 82-3, 86, 94, 112-13, 116, 117, 118, 132, 134, 137, 145-6, 164-5, 181-5, 186, 219
Swanton 195
Sweldon 64-5
Sweyne's Howes 185-6
Swine House 185-6
Sycharth (Denbs.) 32

Taff, river 25, 105
Taibach 23
Tal-y-bont 187
Tal-y-cafn (Caerns.) 77
Tal-y-llyn 187
Tal-y-sarn 187
Talley Abbey (Carms.) 124
Talybolion (Anglesey) 22
Talycynllwyn 187-8
Talyllychau Abbey 124
Tarenni Colliery 225
Tarenni Gleision 224-5
Tarrws 53
Tawe, river 25
Thomas, Hilary M. 14
  R.J. 12, 14, 19, 63, 90, 91, 113, 117, 131, 168, 203
  William 16, 65, 72, 75, 91, 126, 143, 148-9, 166, 167, 170, 199
Threapwaite 199
Threapwood 199
Tinkinswood 208
Tintern (Mon.) 165
Tir Deunaw 132
Tir y Barnard 51
Tir y brin 96
Tir y Bryn 96
Tir y Bwhayen 51
Tir y Byrbais 51
Tir y Dadlau 199
Tir y pumpunt 132
Tir-shet 190-1
Tir-y-pwrs 132
Tirergyd 188-90
Tiryryrfa 31
Tompkin's Wood 208
Ton-y-gof 126
Tongwynlais 221
Tonna 63, 82, 89

Tonpentre 221
Tontrycwal 191-2
Tonyfildre (Brecs.) 153
Tonyrefail 196
Tor-coed 104
Traherne, Sir Cennydd 43
  J.M. 74
Tre-groes 195
Tre-os 200-2
Trecadwgan (Pembs.) 95
Trecastell 195
Tredegar (Mon.) 15
  Ironworks 15
Tredelerch (Mon.) 192-5
Tredodridge 195-6
Tredogan 95
Tref-y-rhyg 196-7
Treferig House 196
Treferig Isha 196
Tregaron (Cards.) 111
Tregarth (Denbs.) 61
Tregawntlo 195, 197-8
Trehwbwb 24, 195, 198-200
Treoda 218
Trerannell 202-4
Tre'rfran 150
Trerhingyll 195
Trevrane 150
Troed-yr-harn (Brecs.) 152
Trothy 214
Tudor Aled 121
Tullagh 145
Tullaherin 145
Tully 145
Tullyrone 145
Tumble 205-6
Tumble (Carms.) 15
Tumble Hill 204
Tumble Inn 206
Tumbledown 204-6
Tumbledown Dick 205-6
Twlc-yr-hwch 185
Twynbwmbegan 206-7
Twyncyn 207-8
Twynywerfa 215
Tŷ Verlon 213
Tylor, Alfred 143
Tylorstown 143
Tymbl, Y (Carms.) 206
Tythegston 106
Tywyn (Merioneth) 56, 87

Uchelolau 209-11
Universal Colliery 28
Uplands 137
Upper Boat 41
Upper Lledrod (Cards.) 224
Upper Race 162-3

Valle Crucis Abbey (Denbs.) 47
Vernel 213
Vershill 213
Verville 212-13
Vishwell 213
Vivod 126
Vurlong 213

Wade-Evans, A.W. 110, 124
Wainsill 186
Watford (Glam.) 19, 213-15
Watford (Herts./Northants.) 213
Waunarlwydd 50, 94, 112
Welsh St Donats 98, 124
Welvord 213
Wentloog 29
Wenvoe 42-3, 53, 175, 188, 209
Weobley (Herefs.) 17
Werfa, Y 215-16
Werfa Ddu, Y 215
Wern, Y 86, 90
Westra 177-9
Wharton Street 216-17
Wheelock, river 13, 225
Whitchurch 217-19
Whitford (Flints.) 215
Whose Land (Wilts.) 199
Wigfair (Denbs.) 103
Wilcrick (Mon.) 225
Wileirog (Cards.) 225
Williams, A.J. 46

Edward: *see* Morganwg, Iolo
Edward G. 215
G.J. 59, 86, 129
Sir Ifor 12, 13, 34, 67, 70, 77, 88, 89,
    111, 112, 113, 118, 126, 142, 171-2,
    173, 188, 224
Womanby Street 99, 183
Wonastow (Mon.) 214
Wormshead 219-21
Wren Castle 35
Wren's Castle 35-6
Wrinstone 188

Yniston 141, 221-2
Ynysawdre 222-4
Ynyscedwyn 21
Ynysweryn 219-21
Ynyswilernyn 224-5
Ynysybwl 114-15
Ynysymaerdy 165
Ystrad 112
Ystrad Eláì 112
Ystrad Fflur (Cards.) 112
Ystrad Marchell (Mont.) 95, 112
Ystrad Mynach 91
Ystrad Tywi 112
Ystradfellte (Brecs.) 19, 112
Ystradgynlais (Brecs.) 88
Ystradowen 56, 112, 130, 142
Ystradyfodwg 112
Ystum Taf 218